Worm Farming

Learn About Vermiculture and Vermicomposting

(The Practical Guide to This Unique Form of Natural Composting)

Gerald Blake

Published By **Zoe Lawson**

Gerald Blake

All Rights Reserved

Worm Farming: Learn About Vermiculture and Vermicomposting(The Practical Guide to This Unique Form of Natural Composting)

ISBN 978-1-77485-808-0

No part of this guidebook shall be reproduced in any form without permission in writing from the publisher except in the case of brief quotations embodied in critical articles or reviews.

Legal & Disclaimer

The information contained in this ebook is not designed to replace or take the place of any form of medicine or professional medical advice. The information in this ebook has been provided for educational & entertainment purposes only.

The information contained in this book has been compiled from sources deemed reliable, and it is accurate to the best of the Author's knowledge; however, the Author cannot guarantee its accuracy and validity and cannot be held liable for any errors or omissions. Changes are periodically made to this book. You must consult your doctor or get professional medical advice before using any of the suggested remedies, techniques, or information in this book.

Upon using the information contained in this book, you agree to hold harmless the Author from and against any damages, costs, and expenses, including any legal fees potentially resulting from the application of any of the information provided by this guide. This disclaimer applies to any damages or injury caused by the use and application, whether directly or indirectly, of any advice or information presented, whether for breach of contract, tort, negligence, personal injury, criminal intent, or under any other cause of action.

You agree to accept all risks of using the information presented inside this book. You need to consult a professional medical practitioner in order to ensure you are both able and healthy enough to participate in this program.

Table Of Contents

Chapter 1: Composting 1

Chapter 2: Worms You Can Utilize 3

Chapter 3: Worm Farming Benefits.......... 5

Chapter 4: How To Design A Worm Farm. 7

Chapter 5: Vermicomposting 9

Chapter 6: Large And Small Scale Worm Farms 11

Chapter 7: Food & Feeding 16

Chapter 8: How The Worm Population Is Managed... 20

Chapter 9: Issues With Worm Farming .. 22

Chapter 10: Other Ideas To Use Compost 25

Chapter 11: Beginning A Worm Farm Business 28

Chapter 12: The Basics Of Worm Farming 61

Chapter 13: Worm Biology: The Anatomy Of An Earthworm 68

Chapter 14: Types O Worms For Vermicomposting 76

Chapter 15: The Basics Of Worm Bins 82

Chapter 16: How To Build A Worm Bin .. 97

Chapter 17: Winterizing & Insulation ... 122

Chapter 18: Feeding Your Worms 133

Chapter 19: Worm Bedding 101 140

Chapter 20: What To Do And What Not To Do When You're Involved In Worm Farming 154

Chapter 21: Worm Bin Care Tips 167

Chapter 22: Troubleshooting 173

Chapter 1: Composting

The composting technique can be used to recycle natural waste. The worms will eat the waste. Once the worms have actually eaten it, they can use it to fertilize their garden. Food waste from fruit and vegetables is best.

The worm compost will allow your fruit and vegetables to grow larger. You'll have fresh food products to eat if you fertilize. If you have a large worm farm you will have lots of compost worms.

Here are some tips for composting.

- Keep your compost piles hydrated. You could endanger it if you add too much water. It must not get wetter than necessary.

- Compost pile aeration means that it is constantly turned on a consistent basis. By doing so, oxygen can enter the compost pile. As long oxygen exists, organic materials are more likely to be destroyed.

Balance of carbon and nitrogen is essential in compost piles. A surplus of either can throw it off-kilter. Balanced compost will consist of a mixture natural material and grass cuttings. Your compost will then flourish.

- This is how composting keeps bugs away and plants healthy. It reduces or eliminates chemical fertilizers. The cost-effective treatment of soil is achieved. All toxic substances and hazardous waste are removed.

It is now possible to compost with worm farms thanks to the "green revolution". It could be beneficial for both children and adults, as well as helping to preserve the environment. You can also consider other aspects, such as having your waste go to landfills or dumps.

Today's environment is contaminated with many chemicals and other substances. It is astonishing that people are still capable of breathing in this mess. You would be surprised at the difference food scraps, and other potential waste might make in someone's lives.

With the help of composting, veggies and plants can enjoy an improved life-span thanks to the worm farm. It will have a significant impact on the environment.

Chapter 2: Worms You Can Utilize

Of course, worms are necessary in order to create a worm-friendly environment. Both earthworms as well as red worms have been recommended for your worm farms. Red worms can be used more effectively than earthworms.

Red wigglers have a faster reproduction rate which makes them more attractive. They provide a lot of nutrients which help improve the soil. It is easier to locate earthworms, but they do not produce effective results if they are left in the soil.

The red ones however have shown to be more capable in producing worms in farms. They can survive different temperatures and can stay in enclosed areas. They could also be used to process large quantities of waste products.

Some worms are better at composting than others. This is why it is so important to choose the right worms for your worm garden. A worm farm could also make use of live bait from worms.

European Nightcrawlers could also be used for worm breeding. This worm is also useful for pets like birds and fish. They can be used for composting but are best used as live bait.

The Red Wigglers, Nightcrawlers, and Red Wigglers could be placed in garden or lawns. They improve soil fertility and function as fertilizers.

There are many types and varieties of worms. However, there are two main types that people use. They are also the easiest to get. You can also find them as young and mature worms. They may be waiting to hatch from the egg. After the eggs hatch, you will discover more worms for each egg.

Your farm will be more effective if it has the right worms. It will surprise you to learn that not all worms are able to compost your waste. There are still others that might help your worm farms stand out.

Chapter 3: Worm Farming Benefits

Worm farming is a great way to make money. Worms are a food source for livestock. If you visit a museum, you might find a worm farm there. Student might make a wormfarm as part of a school project. Worm farms can be used as treats for birds or fish.

A wormfarm could be a great way to teach others how they can recycle their waste. They are also a valuable asset for the trade economy. Worm farms help to keep the soil fertile and healthy by being used in soil. Thus, equipment and products will be required to maintain the farm's natural resources.

Because catfish farms may have their own worm farm, worm farming could also help catfish farms. Fisherman could use worm farm worms over synthetic bait to capture fish. Synthetic bait can also cause contamination in rivers, lakes, ponds. In chicken houses, chickens might use worm farms to provide their food.

It seems that no matter where one turns, there are people or animals who could benefit from a wormfarm. It is an excellent thing because it helps maintain a clean environment. You don't have to deal directly with any contaminants.

Worms are also associated with the vegetables and fruits we eat. The soil which the worms helped to fertilize is where the plants come from. This is a way to keep the soil healthy. Gardeners have used worm farms and worms for years to aid their gardens.

Individuals with an interest or knowledge in worm farm could meet up and exchange ideas. Individuals from different backgrounds could work together to improve the environment. Individuals could meet up in groups to learn more about worm farms or worms.

Other countries have worm-friendly cultures and people eat them as a normal food. It's amazing to realize how much worm farms can actually help the environment.

Chapter 4: How to Design a Worm Farm

As was pointed out quickly at the beginning, you don't have to have a lot of land in order to start a farm. This can be done in your home if you choose. The container does not need a lot of space. You only need enough space to place the container in a safe location.

There are many different ways you can set up a farm. But, it's easiest to start one on the ground. While larger worm farm may require concrete pads, it isn't necessary for smaller ones.

These are the worm farms with larger numbers of worms. Windrows can be worm farms that have been built up into long rows. They produce many of these.

The worms will arrive and start to produce, leaving castings behind in the first row. They will carry on with the second and third rows, respectively. Certain individuals will use wood boards or bricks to keep the worms under control.

Don't forget the cover! Covers are essential for all containers. Worms dislike light. They will produce better results in dark and damp conditions.

You don't have to create your own farm. There are many designs available. There are many different types that can be used in your apartment or house. The basement is where the worm farms are kept if you own a home. But, people have found other places to keep them safe.

As I mentioned earlier, the kitchen location is perfect. If you don't mind storing them under your sink, they can be stored underneath it. You'll be able access the organic and vegetable scraps much faster. This is how you can get started with vermicomposting.

Chapter 5: Vermicomposting

Vermicomposting involves the creation of vermicompost. Vermicompost comes from organic matter created by worms, also known as red wigglers. Vermicompost can be used to create the ideal soil for your worm farms. This is because vermicompost has a lot more nutrients, fertilizing and conditioning than organic matter.

Red Wigglers may be used to make compost. Red Wigglers often come from rich soil found in Europe and North America. They love making soil from compost and manure. You could find these worms at bait shops or order them via mail order.

Properties

Vermicompost has a high level of microbial activity. This allows nutrients to be disintegrated and transformed into plant-friendly forms. You will discover that worm casts have mucus.

The mucus keeps nutrients around, even while watering takes place. Mucus can also hold more water than regular soil. Vermicompost is fast and easy for seedlings and fruit pits.

Vermicompost contains tiny seeds made from kitchen scraps. These include tomatoes and eggplants. These seeds will grow in a matter of weeks.

There are some advantages to using vermicompost and worm farms in your soil. Some of these benefits are:

- Soil enrichment

- A good holder of a specific amount

- Root growth is increased

- The structure has been upgraded

- Enhancement in physical appearance

You could also make worm tea using vermicompost. Use that along with water to allow the mixture to steep for a couple of hours or several days. You can then use the liquid mix as a fertilizer. A home set can also be bought at the hardware store or from the plant nurseries.

Chapter 6: Large and small scale Worm Farms

A vermicomposting box can be used to make a worm farm. Make sure that the bedding has been kept damp so the worms are not trapped in it. If you live in hot and/or warm climates, be sure to keep the bin out of direct sunlight. You can place waste weekly or daily, depending on what you need.

The first time the worms begin to eat, their intake is approximately half of their total weight. After getting used to the feedings, they may eat all of what they weight. The worms have to eat the old food before any new food can be introduced. In the event that you have another bin, you may place the brand new food aside.

Many people set up worm farms in a small area. A collection tray is an excellent idea. You don't want it to spill all over. Plastic bins are more able to drain than wood ones. They are not designed to absorb. The drainage is lower for wood bins.

It is important to consider the location and feeding habits of your worms when you use a small bin. One of the following options could be used to create small bins:

- Vertical flow - Trays are stacked vertically. The first tray that fills up is the one at its bottom. The tray that is full should be kept separate from the one at the top. There will be a layer of dense bedding on top. The tray to the right will have organic matter.

For the worms, the bottom tray serves the same purpose. They can then compost the materials. The worms will then proceed to the next tray. After they have completed the lower tray, vermicompost may be collected. Harvesting is easier when the bins are used in this manner.

- A constant horizontal flow-- The tray are horizontally stacked. This method allows the worms a greater tendency to seek out food and makes it easier for harvesting. The bin is generally larger. A chicken wire screen separates the bin into two parts.

Half of the bins will eventually become full. The other half, which is going to be made up of organic matter and bedding, will then become empty. The food will then be transported to the opposite end of the bin. You could then take up the compost.

- Non-continuous: This arrangement does not use a container that can be split in half. A layer of bedding is located at the bottom. This is used to line the container. After that, organic material is used for composting. Next, another layer of bedding will be added.

After this is complete, composting can begin with organic matter and bedding. This bin is perfect because it is small in size and doesn't take up too much space. You will still need to empty the bin when you are ready to harvest.

People looking to start an enterprise should not build large-scale worm farms. This could be done in two ways. The windrow is more common.

Windrows were mentioned at beginning of the book. They help provide lots and lots of organic matter for the larvae. Wave movement occurs when only one windrow is used for worm feed.

The wave movement starts when the one-sided feed has been completed. There are many castings in the rear part of the wormbed.

A raised wormbed is another large-scale method of vermicomposting. The system could be made flow-through. The worms live on the top of the mattress and consume approximately an inch of the food, which is distributed throughout the surface.

The remaining castings will be collected using a breaker bar that yanks the castings through a large mesh filter. This screen is used to form the base of the wormbed. Red worms will stay on the top of the screen and then gravitate to the new food.

This arrangement ensures that castings as well as worms are not separated before being packaged. This setup is ideal for creating worm farms in warm weather.

You need oxygen so that composting bacteria and worms can work in the bins. Take out any composted materials and make sure the bins have holes.

If this is not possible, you can opt to use bins with constant circulation. Whatever you use to keep the worms happy, they must be able to breathe properly. If oxygen is lacking, you'll notice a strong scent and an environment for the larvae that can be harmful. This could pose a threat.

In order to preserve bins, it is necessary to inspect the bedding for moisture. The oxygen flow must also be checked. This must be done at minimum once or twice per week. It is vital that the moisture level does not exceed the maximum.

The type of bin you use will determine how you need to preserve it. For instance, if your bin is constant flow, you might have to spray water on the bedding in order to keep it moist.

This is because constant flow bins don't contain any liquid. You can release the liquid from a non-continuous flowing bin by turning on a tap. It is not necessary to throw it away. It can be used to make plant food. For those who have compost that has high levels of acid, crushed eggshells can be added to the mix along with veggie matter.

If worms are used in composting, they should be kept at 55 to 70 degrees Fahrenheit. The bedding shouldn't be frozen or heated above 89.6 degrees Fahrenheit. All of these temperatures indicate that indoor vermicomposting is acceptable in your home. If you live in a tropical climate, the only exception to this rule is indoor vermicomposting.

Chapter 7: Food & Feeding

The worms' waste contains a mixture of nitrogen and carbon. To get the best ratio, it needs to be combined and formed. Shredded papers will have more carbon.

You will find more nitrogen in food scraps. You'll also find plenty of protein in food trash. Food waste composed of vegetables and fruit, and which gets rid of any veggies that are high in protein and animal matter, will not have much nitrogen in vermicompost.

There are many food waste products that you could use to feed worms. If you don't use animal matter, you are fine. Worms don't want to be able to digest harmful substances. With the exception for citrus peel, paper filters, plate scrapings and vegetable peels could be used.

These scraps don't need grinding, as the bacteria that is currently in your bin will soften them. You can however speed up the process of composting by cutting up vegetables and fruits. Use a small amount of water to moisten dry foods such as bread that is molded and covered in bedding.

Don't include food waste. A lot of food waste can make the mix unpalatable for the worms. Due to their high protein content, beans should be avoided. It doesn't mean that worms need to consume a lot.

Actually, vegetables and fruit matter are ideal for them. Don't use any grains or food that was already prepared. You should whisk the mixture to ensure sufficient oxygen to the worms. If the bin gets too damp, add more bedding.

Some people use tomato tops and carrot leaves in their blend. You can provide the worms with the nutrients they need to absorb the food waste quickly by adding a little bit of garden soil.

Another thing you shouldn't add is anything that has been sprayed, such as grass clippings. The worms won't be able to breathe if they have to deal with a lot of fat and oil. They are connected to their skin.

Additionally, Worms are not fond of foods that contain a lot of spices and salt. It is best to limit the amounts of foods that contain starch or acids. It would create an imbalance if there was too much.

There are 2 ways you can feed the bin. Top feeding is one. When organic matter is added on top to the bedding, this is known as top feeding.

It is then covered with an additional layer. This is the procedure for each bin feeding. Pocket feeding is another kind of feeding. Pocket feeding involves a top bed layer with food hidden beneath. The food doesn't stay in the exact same spot every time.

The bin rotates, which allows for the food to break down in the pockets that were being fed. The top layer of bedding becomes empty and more food is added.

Both of these methods are combined in vermicomposting. It is important to have bedding material for any materials that may be hidden beneath the bedding. If you don't have enough bedding material, fruit flies may be coming to your home.

When you see no food or bedding left behind, vermicompost will be allowed to collect. There might be obvious items such as peach pits, melon peels or branches.

Gathering can be done in many different ways. It all depends on the intended use of the vermicompost. It also depends on whether you intend to keep a large number of the eggs and worms away from the vermicompost.

Overfeeding the worms would be a problem. This can lead to foul odors or the emergence of insects. You should feed them moderately, providing them with very little food waste.

Do not forget to give them fresh food as soon as the old food has been used up. Worms are tiny and can eat up to half of the food they eat daily. Overfeeding them could cause serious problems in the worm farm.

Worms can create more worms per month. It is vital that you keep track of how many worms are present in order to provide them with enough food scraps.

You can feed your worms well and they will produce plenty of composted material to enrich the soil. This will help the worm and you.

Chapter 8: How the Worm population Is Managed

Three components are used in managing the worm population.

- Food accessibility

- Space requirements

Environment

The worms can be fed every day in very limited space. This allows them to reduce the amount of food they eat and absorb the nutrients. They dispose of anything they don't eat. More worms fight for food as they reproduce.

The worms will become more numerous and require more food. Additionally, they will need more space so that their bedding can be kept fresh.

All worms force castings out their system. Castings is mucus which contains little grain particles. Castings become harder when they are left open to the elements. Even with all that, mucus is not able to break down or seperate easily.

This is what causes the soil to become more compact. This allows the drainage system to function more efficiently. You can see that the castings work as both a soil conditioner and a fertilizer.

The purpose of having many worms is to increase the population. You will need to collect vermicompost at a slightly lower level if you have plenty of worms. Otherwise, the castings can become poisonous. It is easier for worms to move away from the castings than it is for others. In fact, many worms may live for longer periods of time.

Chapter 9: Issues with Worm Farming

There are many possible reasons your worm farm might not be working properly.

Bugs

It is important to keep bugs and other insects out of your container. It will keep a lot more bugs out of the worm farm by preventing them from contaminating it. There are still going to be some bugs that try to get into the worm farm. The vinegar flies are usually not seen by those who produce worm farm. This is because the worms are being fed too much. The flies will grow larger when there is lots of food.

Ants are known to have an inclination to become another problem. This indicates that the soil could be too dry. To add moisture to the soil, you can pour some water. This will aid in the elimination of ants. Vaseline can also be applied to legs of worm farms to keep them from climbing up.

Maggots are likely to appear in your meat. The worms should be evicted from any meat you have. In the worm farm you could place a little bit of bread that has been soaked in milk to kill maggots. Maggots will stick to bread and you can get rid.

Here are some other methods to prevent ants and other pests from polluting a worm farm.

- You can add a bit lime and water to any worm farms that have too much acid.

- Keep your worms safe from drowning in a worm farm if there is a lot of rainfall. To be able to stand straight they will need to be at or near the top of their worm farm. Another thing you could do is to absorb the excess is to include paper and cardboard.

Wrapping the additional food in newspaper is another option for a foulworm farm. Keep it in a cool spot or in the refrigerator for a while.

A Hessian blanket is another way to cool worms in hot conditions.

- For worms to reproduce, it is essential that they remain in the sun.

These tips and others could help you to keep your worms happy and produce plenty of soil for the worm farm.

If you notice that worms are trying to leave, it may indicate that they don't want to stay where they are. One reason they might be leaving is dry soil or lack of food. Sometimes the soil is too dry and the worms feel uncomfortable. It doesn't matter what, the issue must be fixed immediately.

Water should be added to the worm farms if the soil is dry. Get rid of any excess soil wetness and replace the farm bedding with fresh bedding. It is important to determine the source of additional wetness and to get rid of it.

Assure that there is plenty of food available for the worms, and that the temperature is comfortable.

Chapter 10: Other Ideas to Use Compost

Compost Garden

Composted compost could be used for many other purposes than just a wormfarm. The compost is currently used as a fertilizer to plant and improve soil conditions in gardens. It will drain well when the soil is healthy. Soil that has been enriched with compost will retain its drainage and texture.

You will find many nutrients in the compost. That will make it easier to keep your plants healthy. The fact that plants don't require too much nutrients at once means they can't get them all at once. They prefer to have small amounts of nutrients.

Compost keeps certain materials from going to landfills. The same materials can be used to make your worm farm. They include coffee grounds as well as cardboard, vegetable peelings and egg cartons.

Please be aware that there are certain things that must not go in the compost. Meat and animals should not be added to the compost. It could pollute the compost making it inedible.

If you have a suitable place, you could put the compost stack on the soil. Keep it away from the sun and moisture. Continue to add things to the piles. This will allow it to grow. It might take some while, and the compost that is done will be on top of the piles.

A shovel can be used to hold everything together. This will accelerate the decay process and preserve the items together. You could put the compost into a container. You could also use an empty trash bin to add holes. You could even buy a new bin for your mixture. Rotating the mixture should be done several times. It can be done often, no matter how many additional items are added.

You will be happy that you used this natural method to care for your plants.

Compost Tea

You should know that this is not your tea. Do not let the colour fool you. Follow these instructions while making this specific type of tea.

1. Check out the compost to see how you can use it. You can also use materials that have not been

completely composted. Take a few and take it out of the stack.

2. You can fill a bag made from fabric with compost. You should keep the bag no larger than the medium. You need to be able and confident to carry it.

3. Place the bag into a container that is large enough to hold it. Let the bag sit for 3 days.

4. To ensure the water does not go dark brown, add more clear water. This mixture can be used to plant.

5. It is possible to reuse the same compost bag several times.

6. You'll learn how to stop using your bag when the water becomes dark brown. You won't get any more nutrients.

Keep in mind that compost tea can be used for both plants in pots and those in gardens. It could also be used for plants that grow outside and inside.

You could also buy a pump at a home aquarium. Make bubbles with the hose.

Chapter 11: Beginning A Worm Farm Business

Do you have any thoughts about starting a worm farm company? It is possible to start your own worm farm business, or even create a small side venture. It is essential that you have the knowledge and skills to run your own worm farm. While there may be problems after the fact, you shouldn't be worried as long as everything is clear and simple.

Before you embark on this journey, there are certain things you will need to consider. Be sure to have the proper bedding, containers, and food waste materials. The climate of your home is another important thing to consider. It is crucial that the environment where the worms live is comfortable, but not too cold or too hot.

Additionally, vermiculture will be of interest to you. Vermiculture means that you use worms as a way to accelerate the process of composting. It doesn't take up much space, and it is very odorless.

Before you start to handle worm farms on a larger scale, make sure that you have as much experience with this process as possible.

Prospective customers should see that you have a solid understanding of what you are doing.

Now that you have perfected the worm farming process, it is possible to start a business. You'll need to have sufficient materials to supply your requirements if you want to grow worm castings and castings.

To do so, you need to locate suppliers who could supply your needs for materials and other products. You will have a more successful business if you are able to keep your materials in stock.

A valid business license is required to launch a company anywhere in the world. For the sale of worms and other accessories, you may need to obtain a separate license.

Once you have established the inventory, you will require customers. Even though you may know that your products and materials are excellent, the potential customers might not be aware of this. You'll need to be out on the street to find customers.

There's no doubt that you will need to invest money in marketing. You must ensure that marketing works well and that you're getting the best value for your money. One way to get word of mouth started is by building relationships.

This kind of marketing has always been one of the most popular ways to spread the word. This can only take you so far. It is necessary to diversify your efforts to spread the word and promote the product.

If you are just starting out, it might not be financially feasible to pay for radio advertisements. Similar reasoning could be applied to magazines and newspapers. Take a look at the cash you have available to pay for these expenses. After everything is said and done marketing expenses add up.

One of the best ways to market your business as a worm farmer is to create leaflets. It is possible to make several stacks of leaflets and then distribute them. It is important to plan where you want them to go. It is simple to put them under your windscreen wipers. But, is this the best way for you to maximize your cash?

There's also the risk of injury. What happens if your flyer is stuck beneath the windscreen wiper and causes damage? Now, you have to change that part of your wiper. It might not be as effective. Additionally, you might find 10 people who are interested in what it is that you can offer out of the many cars that are parked.

If someone saw you place a leaflet in their car, they might think you're trying to steal it. Laundromats or supermarkets are good options for leaflets. Before you send your leaflets, be sure to get the approval of the owners. They may have a no invitation policy.

Are you thinking of giving a talk on worm farming to your local library? Children might be interested to start one. Many children love to take care of the natural environment. This would be a fantastic job. It is possible to create bookmarks that include your business information so it isn't obvious what you are promoting.

An advertising sign for your worm farm can be also made. By making the sign bright and easy to see, you can make it stand out. It should look great so that people see it immediately as they pass.

A larger sign would do better than one that is smaller, since people will be more likely to notice it. A plain sign won't do much for your business. You want something that will say "Let me take a peek at this place!"

You can also use signs that are posted on the doors of vehicles. These signs are more visible when you are outside. This is another cost-effective way of getting your point across. You can also place online ads on Craigslist and other online advertising platforms.

You can promote your products in newspapers by starting with the ones that are most inexpensive at restaurants and other places.

You could also set up a stand at a local farmer's and similar markets. Renting the area would mean that you would need to pay an additional fee. Make sure you have plenty of materials on hand.

It is essential that you continue marketing your business. There are only so many customers that you can keep up with. You want your business grow. To do that, you have to keep marketing to more customers.

Referral programs are a great way to do this. If your clients refer others to your business, you would offer a discount. This is a good thing to do if you already have a large customer base. Profits are essential for any business.

Do not believe any supply companies that claim they can make quick money from worm farming. It is just a process like any other business. While you will ultimately make profits, you shouldn't view it as easy cash. For some businesses, it takes longer to get off the ground. You can take pride in your work and the sales will follow.

A word about vermicomposting

Vermicompost can be used to refer to red worm-compost products. It is the most nutritious fertilizer made of compost available and is 100% organic.

How do You Make Vermicompostable?

Red worms are the best choice if your goal is to become a vermicomposter. The red wiggler is only one species that can produce vermicompost. It is important that your worm farm has adequate moisture.

Vermicompost is a great option

* It adds variety to the soil's appearance

* The soil will be packed with additional nutrients

* The only type of worm capable of making vermicompost is the redworm. Redworm castings contain a mucus to protect the soil from the harmful effects of the soil.

What is so special about vermicompost?

Vermicompost can be made from microbial organisms. They help to degrade extra nutrients in soil. This mucus comes via the redwiggler's castings and acts as a protector against moisture and other seasonal decompositions of nutrients.

Vermicompost does a great job. You can even get some natural yields from kitchen scraps when you add them to this type worm farm!

This potency will continue in soil as well as if vermicompost is used to make your worm tea. This "tea" is a potent liquid fertilizer that can be steeped in water for up to seven days. These are just three reasons why vermicomposting works so well for worm farms.

Helpful Hint. Red worms do their best work in soil production when they have access to manure, tomatoes or eggplants.

How to get started with Worm Farming.

* Worms (earthworms, red worms) will suffice

* Bed bedding for the worms (this can be made from cardboard or newspaper, and even a shop-bought one)

*Bedding is crucial for worms, as this is where they obtain their nutrition. It must include one of the listed materials (hay, burlaps, sawdust, aged and dried manure leaf, or hay), and it should be set up so the worms are able to move through the bedding.

NOTE: You should avoid using glossy, wax and copy papers with recycled paper. These paper types can contain substances that can harm your worms ability in the composting process.

* Housing for theworms - specifically, a wooden container or plastic one**

**Some people suggest only using metal containers. But metal can be a compost contaminator which can lead to the destruction of your worm farm.

* Organic waste is vegetable scraps or tea bags. ***

***Do not mix animal waste with your Worms. It could contain germs or other pestlike substances that can cause harm to your Worms.

Worm farming is very affordable once you have the supplies in place. Let's begin with bedding. Worm bedding can include straw, compost, and even newspaper. These pieces should be cut into 1- to 2-inch pieces. The time required to pick up the recycling container will probably be less expensive than the materials. Another example is housing. You can buy pre-made worm farms containers or you can make them yourself by drilling holes in transparent plastic containers.

Why use a see-through container? So You can enjoy the sight of your new friends!

You will need at minimum two good worm container, one to house the worms, and the other to store any worm castings.

How much space do I need?

Worm farming doesn't require much space. A worm farm can be done in an apartment provided you have enough room for the containers. In a later chapter we will talk about how the containers should be laid out. The bottom line is that you don't need a lot of space for worm farming.

Different kinds of Worms for Farming

Before you start worm farm, you must choose which type ofworms you want to use. The three types of worms that are most commonly used by worm farmers are earthworms, nightcrawlers and red wigglers (aka, red worms). This guide will be discussing worm farming using red worms or nightcrawlers. These are thought to be more easy to obtain and easier to work with.

Useful hint: Red wigglers are available online or in bait shops.

What makes red worms & nightcrawlers so wonderful? Take a look at the following facts about each one:

* Red worms and nightcrawlers are better at composting earthworms.

* Nightcrawlers are the best bait around

Red worms are known for their ease of reproduction, even when they live in enclosed areas and at varying temperatures.

* Red worms or nightcrawlers may be released into gardens. There they will happily enrich the soil and help it thrive.

Red worms, and nightcrawlers, are easy to find because they can be obtained at their egg stage.

Your purpose for worm farming will dictate the type of farm worm you choose. For raising fish bait, nightcrawlers could be the best. However, if your goal is to keep it as a hobby, you might consider red worms. The best choice will ultimately be the one that serves your purpose, whether you are looking to raise fish bait, for farming business, or just for enjoyment.

A Layout for Your Worm Farm

There are many different ways that you can set up your worm garden, depending on what appeals to you. Worm farms can vary in size depending on their purpose. Many people who run worm farming businesses use a windrow based setup. This means that the worms are spread out in a series of rows. However, a smaller worm farming operation (maybe set up as a hobby) might work in a much smaller space than a cupboard. The purpose of your farm, the supplies you have available, and your personal preferences will determine the design that you choose.

Helpful Hint: Some people worm farm on concrete. This is unnecessary.

What guidelines should you follow to start a farm of worms?

* Worm farms should not be stored in bright areas. Worms are more productive in darker places. You should cover the space where you store the worms if it is too bright.

* Basements or kitchens can be ideal for setting up worm farms.

* You should look for a space where you can store a container.

* Premade worm farms can be a great option for anyone who is having trouble choosing a design.

No matter if your goal is to start a business in worm farming or just as a hobby, planning ahead will help you ensure that your worms produce in an efficient manner.

Worm Farm Layouts

You can get a little setup for as low as...

Vertically Divided Layout. In this layout, the trays are placed on top of each other.

It looks like:

* Organic supplies

* Betting (done thickly).

* Worms

* Worms and compost

The following layout offers benefits:

* It is very easy to take out tray after tray to harvest.

* Worms can travel between layers to make composting more efficient.

Horizontally Divided Layout. This method places the trays horizontally instead of vertically. The bin can be divided in half using chicken wire. Fill one side with bedding for the worms or organic supplies and fill the other with food.

The following layout offers benefits:

* It's possible to harvest compost less often, for example, when the trays become half full and not waiting until they are full.

Layout for the Bin that isn't Divided - This layout does not allow the bin to be divided.

Looks like:

* Extra bedding

* Liners of bedding

* Worms, organic products and compost

The following layout offers benefits:

* Takes up less space

Notable Note About This Layout: The bin isn't divided so you will always need to empty it out when you need supplies or castings.

For A Large Setup...

Remember that worm farming can be a lucrative business venture. We'll be discussing two common methods for large-scale, worm farming.

Raised Worm Bed- This setup has multiple layers within the container. Each layer is responsible for caring for the worms.

It looks like:

* 1 inch Layer of food

* Mesh screen is used for bedding and casting.

The following layout offers benefits:

* No need to separate worms from castings

* Works well in warm climates

Windrows – This arrangement places the bin in rows that each have a specific purpose for caring for theworms.

Looks like:

* Feeding section

* Section Bedding

* Castings Section

The following layout offers benefits:

* Easy to clean without needing to remove the bugs

Castings are very easy to remove, as they come with their own section.

* The most preferred layout for large-scale worm farming

Air: A Word about It

Signs that there is an issue with air flow

* Unexpected deaths of worms

* Odours are released from the bins

Special Note for All Layouts on Air Flow: If your containers feel suffocating, you can try to create holes within them to increase airflow.

No matter the layout, it doesn't really matter how you design your worm farm. The key to creating a worm farming layout is making sure that you make it as easy as you can for them to care for and that they have everything they need (food and water, humidity, food, etc.).

While there may be occasions when adjustments will be necessary, it is important to recognize those times. You should ensure that the worms receive the correct amount and quality of water and air. These things will not only be necessary for the food and bedding but also the overall process.

Things You can do to adjust the levels of moisture and air

Let's now consider how you can help your worms thrive.

* Add extra holes to the container for better air flow or adequate drainage.

Sprinkle water frequently over areas that lack moisture (to encourage adequate moisture).

* Watch the temperature! Worms are most at home between 55 and 70 degrees Fahrenheit. Anything lower or higher could be harmful to your worms.

If you have any questions about the different layouts of the book, or if you're planning your own layout, feel free to refer to this chapter.

Feeding

What Do They Eat

Now that you have set up your worm farm, we can get into basic care and feeding. They are great composters and can eat most things. Below is a list with the only things you should be cautious about when feeding your Worms. For your worms to be healthy, avoid adding these ingredients to the mix.

Foods to Avoid

* Meat from animals by-products

* Grains, when possible

* Oily or fat foods

* Spices

* Anything that has come in direct contact with pesticide (this also includes grass)

Avoiding the food items listed above is important for two reasons: imbalances and contamination.

All the food items listed here are either carriers of contaminants/animal by-products/meat or can tamper your worms digestive system. The worms will not be able to digest some of these foods, such spices. This is true for other foods as well, like grains. The grains absorb the water the worms need to live in their home.

Note Concerning Dry Scraps

Don't feed your worms too many dry foods like bread, cereals and other dried goods. Dry foods will absorb some moisture, which can cause problems in the balance of your container. This can help to counteract the drying effects.

If you have trouble controlling the moisture, soak the scraps or cover it with bedding. Balance is the key to food, just like with many other worms.

A balanced diet is the best way to eat.

We have discussed the possibility to feed your worms any leftover food after a meal. This is true, and will work, provided that the scraps of food you have left after a meal are balanced with shredded newspaper. This is because shredded newspaper is high in carbon while leftover scraps are high levels of nitrogen.

The best diet for worms is one that contains both carbon and nitrous. Incredibly, too much food scraps can result in an unbalanced diet that can be more harmful than not feeding them at any time. Excessive nitrogen can lead to acid, which can be fatal for your worms. Therefore, it is important to have a balance between shredded paper and food scraps.

Don't include scraps that were obtained from an animal. Animal products can be a primary carrier of bacteria and contaminants that could harm your worms. If you don't, your worms may become sick.

Preparing food for the Worms

There aren't many things you can do to prepare your food. To make it more accessible, you can cut up the paper and chop up the scraps. This will help speed up the composting process. The most important thing is to give the proper amount of carbon to your worms.

Feed Your Worms

The guide will cover top and pockets feedings.

Top Feeding

Top feeding is a term used to describe the method by which you put food on top, then add more bedding.

Pocket Feeding

Pocket feeding is, as its name suggests. This involves storing the food in smaller sections of bedding so that it doesn't shift or get buried. The food pockets in the worm home are then covered with extra bedding. Each pocket is then rotated semi-regularly.

So which do you prefer, pocket feeding or top? Both. Experts recommend using both. Top feeding is an effective method of preventing fruit flies from entering the container. However, pocket feeding can be more beneficial.

How Often should Worms Be Fed

Worms should not be fed until the last feeding has been completed/has been taken down. Because worms tend to be extremely gluttonous, and can consume several times their body weight when allowed. This can make worms susceptible to overfeeding. To avoid this, you should only feed worms their old food.

A Note on New Worms. You need to closely monitor when your worms eat, but you should also look for new mouths to feed. Worms reproduce often (every few month), so it is important to ensure that all worms are being fed in the container.

How does this fit into my Vermicomposting Plan

If you want to vermicompost, it is possible to harvest the results when the food that was put in is largely gone. There are several ways to collect vermicompost. It all depends on what you did in the past.

Conclusions

* Worms must have a balanced diet consisting of carbon and nitrogen. Mixing shredded paper with leftover food scraps can help achieve this.

* Keep worms from eating scraps. You should also monitor the moisture levels of the food to make sure it is not altered.

* Do not feed worms if they haven't eaten the food that was given to them. This will avoid them overindulging.

* Track how many worms your farm has at any given time. You will be able to tell when to start feeding more.

Worm Farms: What are the Common Issues?

It is normal for issues to occasionally arise on worm farms. They can be managed by being vigilant and keeping track of any issues that may arise. This chapter will talk about the most common problems that you might encounter with your worms as well as how to address them.

Acidic Soil

There will be a lot of suffering for your worms if there is an imbalance in the acidity within your worm farm. For acid control, you can add lime and water to reduce or increase the amount of food scraps.

Bugs

There are many different names for insects, including pests and irritants. But if they are found in your worm farm, you know you have a problem.

How to recognize when you have bugs

When you start to see them in the container you'll know they have bugs. Pay attention what kind of bug you are seeing (small, big, or ant). This information will be useful later in determining the best course of action.

Here are some ways to avoid bugs

* Ensuring you close your container lid properly

* Be careful when feeding the soil.

* Monitoring the moisture levels within your worm farm

What to do once you get bugs from the Farm

Once you spot bugs, the best course of action is determined by what type of bug it is.

The soil lacks enough moisture to support ants. Increase the moisture level by monitoring your soil more carefully. You can also move the container into a difficult area for the ants.

If you're seeing maggots, you probably haven't strayed too far from the safe to eat list. Put a piece if bread in some milk, and then place it in the worm farm to trap maggots. Do not feed your worms meat or animal products.

Vinegar flies can be caused by overfeeding. This can be solved by temporarily decreasing the amount of food or monitoring how often your worms eat.

Drainage

If water is not draining properly, you can poke additional holes in the container and add cardboard or paper to soak up any excess moisture.

Runaway Worms

If your worms are persistently trying to escape, you should inspect the environment. Or else, the worms might want to stay.

Your moisture levels should be checked, as well as your temperature and whether your worms are receiving enough food. You should not have to worry about worms trying escaping if you are aware of any imbalances.

Remember that worms are dependent on air, moisture and food.

Smell

Additional paper can be used if your worm farms are odourous.

Temperature

Give your worms extra time in the sun to help regulate their temperature. If your problem persists, try using a Hessian blanket.

Conclusions

It is normal for worm farms to experience problems, especially if this is your first hobby. The key to a healthy worm farm is to pay attention to it so you can spot any problems and correct them immediately. This will ensure your farm produces and keeps your worms healthy!

Keep Your Worm Numbers in Control

Worms reproduce frequently. This process will continue as long you provide support for the existing population. What is the best way to do this? Keep track of how many worms are in your worm garden and add food or bedding to match.

Why would you want worms more? More worms means more castings, which leads to fertile soil. Castings are important because they increase the soil's ability to retain moisture and fertilize it. To increase the soil's fertility, many people encourage their castor population to grow.

Castings: To prevent castings from making the worms environment toxic, you need to get rid of extra castings and worms sooner.

You should remember that worms all require the same basic necessities: food, shelter, water, and moisture. As the worm population rises, you need to ensure that these needs are met. If you care for their food, bedding, space and other needs, you will have more worms.

Running a successful worm farm

Worm farming is a fun and effective way to make your environment more sustainable, enrich your soil naturally, or just to have fun. It is an organic way to fertilize your plants at a very affordable cost. You can also get into composting by using this system. This will help you reduce your garbage output, and also save you money on supporting the worms. This hobby can save you money, benefit the environment, as well as be fun.

This chapter will go over some of the tricks we've discussed to ensure a worm-farm runs smoothly.

Tips to run a successful worm farm

* Before putting the worms into your farm with everything they need to stay healthy and productive, always ensure it is well-set up

* Feed your worms often, but don't give them too much.

* Any leftover scraps should be placed in bedding, then covered with another layer of bedsheets to maximize their consumption.

* Make sure there is enough moisture. You can add more bedding to make it more humid, or you can drill more holes in the container if it drains too slowly.

* Check in often and monitor the farm's health, paying attention to signs that something might need to be adjusted.

* Take note of any new worms. If necessary, adjust your food output and bedding. To house more worms than you can fit in the current container, you might consider setting up a second farm. It's not true that you can't grow!

To enrich the soil, harvest half the castings at once.

* Harvest surplus worms for breeding and/or sale to suppliers in your area.

* Have your kids get involved for family fun!

Take these tips into consideration and the information provided in this document to ensure your success.

The Worm Farming Business

This guide has already helped you get to the point where your worm farm is thriving. Now, what's the next step? This chapter will look at the possibility that worm farming could become a lucrative business instead of being an enjoyable hobby.

The best part about worm farming is the fact that it can be done on a part-time basis. Even though worms will not take up most of your day, you can still make money from them. Although this does not mean that caring for worms is easy, you can still make a profit by doing your research.).

What is the best way to enter Business?

* Proper supplies. This means that you have enough food, bedding, and somewhere to house your worms. Before you can make your worm farm a successful business venture, it is essential that you ensure they are producing healthy worms.

* A good knowledge - Are there any vermicomposting basics? Are you aware of what to do if there are bugs? Is there anything you can monetize from this? This will save you time, stress, and money later.

* A good level of experience - If you're new to worm farming, I wouldn't advise starting a business immediately. You should enjoy your new hobby. Give it a chance, then take some time to evaluate how it goes and decide if you like it. It is important that you have sufficient experience in the field so customers can trust you.

* Some marketing skills - It is possible to be a great farmer but not have any success in business. It is important to have a plan in place to attract customers. We will discuss this further below.

* There should be room for expansion. This could mean expanding on a larger scale or securing connections with other suppliers. It could also mean selling your garage. This is why you will need to have both space and experience.

Additionally, you should research the local regulations regarding what licensing, educational or experiential requirements you must meet, as well as registration. It varies from one area to another, so you will need to research.

A few Notes About Advertising

Advertising can be as simple or as complicated as word of mouth and flyers in the newspaper. Consider how much you would like to invest before starting a business. This will save you the financial stress later when you have to figure it all out.

Let's examine the various forms of advertising:

* Commercials can be used on radio, television or in newspapers. But be aware that it is the most expensive form of advertising.

Flyers can be cost-effective but must be planned well in advance. How do you distribute the flyers.

You don't want to leave the flyers under windshield wipers. Flyer distribution is best done at your local grocery store, discount store, or community bulletin board. These are places where people go to find information and check out flyers. It is a good idea to speak with the owners before using flyers.

* Referral and customer loyalty programs - These programs are becoming more popular. They keep people coming back and want to talk about it. Even though they may only get a modest discount for referring business or repeat customers, it's still a great deal that no one else has.

* Signs-Simple, yet elegant. A sign that is big enough and bright enough to represent your business's image outside may be a good idea.

* Word of Mouth - This is the cheapest and most effective form advertising. This is the most cost-effective method to advertise your business. It is highly recommended but should not be your only source.

* Education - Why not teach others about your hobby. Go to your local library to offer to teach a seminar, or to help create information on a

website to assist potential worm farm owners. This is another way of spreading word of mouth, but also gives you the opportunity to discuss your new hobby or business with people who are already interested.

Advertising will not only cost you money but also bring in new clients. To be successful, every business will have to pay this expense. You should consider this before you open your business.

Conclusions

While the worm farming industry isn't right for everyone, it can provide a great opportunity for those who are serious about worm farming. Make sure you plan out your budget and look into local regulations. You'll be on your way to succeeding.

Chapter 12: The Basics of Worm Farming

Vermicomposting also known as worm gardening is the use of worms to create vermicompost. The worms eat food scraps and any organic matter that we feed. Vermicompost may be harvested and used for soil improvement. This provides essential nutrients for growing plants, improving their quality of living.

If you have all the necessary supplies, vermiculture is very easy. If you have a larger farm, maintenance and upkeep will be necessary. A small space is not necessary. There are numerous benefits that can impact your life as well as the ecosystem. Before we go too in depth, let me briefly describe worm farming and the history of worms.

A Brief History Of Worm Farming

Worms are an old species that dates back as far as the dinosaur age. They have been around much longer than humans. This is what I found out: between 51-30 B.C., Cleopatra VII the Queen of Egypt discovered that the soil was a crucial part of fertilization of Nile. The queen realized how valuable the worms were and banned their export. The penalty for the crime was death. This, it is believed, is why the Nile is said to have the most fertile soil.

Charles Darwin later claimed that earthworms had been one of the greatest creatures on the planet during the 1800s. His research and discoveries were abandoned when the Industrial Revolution took place. The Industrial Revolution was the birth of chemistry. Before you know it chemical fertilizers were a booming industry and the depletion and destruction of soil and earthworms had begun.

"Worm farming" became very popular in the past 40 years. It's still relatively recent. Mary Appelhof, a renowned teacher in vermicomposting, started production of her earliest version "Worms eat My Garbage" in the '70s. In 1980, she hosted her first vermicomposting seminar.

The term vermicomposting was thought to be synonymous with "scam" at the time. But they kept going into the 1990s, when the industry's "Glory Days". Vermiculture is back in fashion, although it did not disappear. It was once a well-known topic in certain circles. Now, it is a hot topic everywhere!

Why should you have a worm farm

There are many reasons that worms can be raised. Although they may seem small, worms have an important role to play in all aspects life. Without healthy earthworms we would have very little food to eat, few trees to give us oxygen, and a limited supply of building materials. Life might look more like an apocalyptic film than it is now. Worms can improve soil health, make food for other species, and provide bait to help us catch food.

Worms improve the soil and plant life

To make soil rich and healthy, worms must be present. While worms live their best lives in the soil, they can alter the soil's structure and nutrient dynamics as well as plant growth and water movement. Some people believe that worms shouldn't be necessary for healthy soil systems. However, their presence is an indicator of a healthy soil system. I would say that worms are essential for healthy soil systems. If they're not present, the environment can gradually become toxic or unhealthy.

Earthworms Enhance Microbial Activity In Soil

A higher level of microbial activity in soil promotes the rapid cycling of nutrients made from organic matter. Then, it is converted into a form which plants can readily absorb. It helps speed up the process for providing nourishment to plants.

Plants use some energy to find the nutrients they need in the soil. Plants can become stunted and unable to get enough vitamins and minerals. The more roots can find food easily, the less energy they need to be fed. This makes it possible to produce beautiful flowers and delicious produce or both.

Mixing and Aggregating Soil

Earthworms excrete waste and poop as they move underground, eating soil. Casts are a kind of soil aggregate. Charles Darwin spent much time researching worms. In his calculations, he found that earthworms can move large quantities of soil. They can flip the top six inches or more of soil in as little as ten to twenty year.

It is essential that soil aggregates protect the soil against wind and water erosion. A stable soil aggregate increases porosity and the retention of nutrients. A bag of aggregate 84.9 Liters can cost anywhere from $35 to $45. This is 3 cubic feet of aggregate!

Enhancing Porosity

Some earthworms can make permanent burrows in the soil. These burrows are often left empty after the worms die or move on. It increases soil porosity as the worms move through it, which can facilitate soil drainage under heavy rainfalls. The burrows can still minimize surface water loss, so it's a double win.

The water-holding ability or retention of earthworms is improved. Gardeners will benefit from the soil's increased propensity for aggregation and greater water holding capacity. The burrow holes, which are porous, are immediately filled with all the available nutrients. This makes it much easier for roots to grow. It is the roots' ability to absorb nutrients faster than usual, as I am sure I have already said. The aerated soil allows the roots to penetrate deeper into ground.

Side hustles or small businesses are great opportunities

People visit bait shops all the time to get worms. But not everyone who buys worms will go fishing with them. Some take the worms home to start a farm. As the farm grows, you can also sell the worms to make a little extra money.

I mentioned before how expensive bags full of soil aggregate can run. Once you've bred enough worms, you can start making some money by selling some. You can keep some castings for your own garden. Castings can also be used for potted plants even if your garden is not in the ground.

Also, pet shops sell worms. It is not bait but food for other creatures. Surprised at how many animals eat them? Rats, birds, snakes (skunks), frogs and salamanders to name just a few.

For the Environmental and Ecological Benefits

Everything earthworms do to the soil benefits the whole environment. An increase in nutrient availability and improved drainage all contribute to the planet's overall health. They are essential to the health of our planet.

As bait or food

Perhaps you are a frequent fisherman or you keep your backyard birds and lizards. The cost of bait can be cut by farming your own worms. A worm farm is a great way to save time hunting worms. If you're like many people, and you catch your own fishbait, then you won't need a worm farm unless it is something you enjoy doing.

A few of the animals that eat worms include chickens. You can find a variety of worms in pet shops. These range from super worms (mealworms) to red wigglers (wax worms). You can also provide worms for animals that require them.

Chapter 13: Worm Biology: The Anatomy Of An Earthworm

It is always a good idea understand the biology of animals and creatures you keep. I love science as well, so it could be me.

Annelida and Annelids are biology and science fans' favorite phylum for earthworms. Latin translation of the word means "little rings". An earthworm's body can be divided into about 100 to 150 segments.

The segments of the earthworm allow it to move on the earth. Each piece or section is equipped with bristles and muscles called "satae". The satae serve to anchor and control the crawler as it moves through the soil. The bristles keep a part of the worm in place while the rest move forward. An earthworm's sections can expand and contract independently. This allows for the body's lengthening in one section while contracting at another. Segmentation plays a key role in worm flexibility and strength. An earthworm would not be stationary if all its segments moved together instead of separately.

Interesting facts

Earthworms aren't equipped with eyes or ears. Even without eyes and ears, worms are able to sense vibrations instead of sound. They also have light receptors which let them know if it's dark or light.

Other vital organs that worms lack are their lungs. A worm does not need to have lung because they can breathe through their skin and mucus. While worms aren't able to have hearts, they do possess a structure similar to a heart called the aortic arch. In fact, worms have five arches of aortic arch throughout their body.

Though worms can have brains but they aren't very complex, Their tiny brains are attached to the skin and muscles of the worms. A worm's brain controls their feelings and movements.

The Circulatory system

The circulatory system in worms is just as important. Earthworms have closed circulatory mechanisms, so their blood flows only through their vessels. Three main vessels supply blood from the worm to its organs. These vessels are the ventral, aortic arches as well as dorsal and aortic blood vessels.

A worm's five pairs of aortic arches functions like a human's hearts. Earthworms have five pairs or aortic arches which pump blood into their ventral and dorsal blood vessel. The ventral vessels bring blood to the back of the earthworm's body from the ventral vessels.

Respiratory Systems

Earthworms breathe through skin, so they don't need lungs. A process called diffusion allows carbon dioxide and other gases to pass through the skin. For diffusion to occur, the worm's skin must be moisturized. Through the release of body fluids or mucus, worms keep their skin moist. This is why they live in moist conditions.

The temperature is lower at night, which is why earthworms are attracted to the surface. The sun, wind and quality of the atmosphere can affect worms' respiratory systems. This is because the moisture in the air will also impact the worms' skin.

Through their receptors, worms can detect light and shadow. An earthworm's photoreceptor, a piece of tissue in its head, is sensitive to light. This tissue allows the earthworm to sense light without having to see it.

The Digestive System

Surprised to find out that the earthworm has a divided digestive system? Each region has its own function.

Pharynx

Esophagus

Crop

Intestine

Gizzard

Worm food travels through the mouth, where it is swallowed by pharynx. This food passes through the stomach and is swallowed by the pharynx. Calcium carbonate is released from the calciferous glands, which rids the worm of its excess calcium. It then moves into the crop and hangs there for a while, before finally making its way to the gizzard. The gizzard of the worm and its crop work just like chickens. The food is thoroughly ground by the gizzard using tiny pieces of stone.

Once the gizzard is done, the food goes to the intestinales. Here, digestive fluids can be released by the gland cells. As with our digestive process the food is digested fully and the nutrients are passed to the blood cells.

Reproductive Systems

Earthworms are hermaphrodites because they have both male sex organs and female ones. Both the male sex organs can produce eggs, sperm, and both can be produced by female organs. It is important to note that other worms can also reproduce. After the eggs are laid, baby Earthworms will hatch in approximately three weeks.

The Life Cycle for an Earthworm

Earth's farmers, also known as earthworms, have three stages in their lives. They are born, live for a long time, then die. Baby earthworms begin life in fertilized eggs inside a cocoon.

It takes only a few weeks for worms to become sexually mature enough to reproduce. About six months after their first sexual maturity, they live to full maturity and will continue living until they die. For most species of earthworms, life cycles are similar. However, there will always been exceptions.

Conception & Early Life

The earthworm's lifecycle begins at conception. Two worms will mate and the sperm is kept in sacs. Once the worms have conceived, they begin creating cocoons which contain fertilized eggs.

Once the cocoons are ready the earthworm will move eggs into them and fertilize with the sperm obtained from the other. The process of cocooning continues until there's no more sperm. They place all their eggs in a plastic basket and drop them into the ground. The baby earthworms will start hatching within the next two-three weeks. One to five worms usually hatch from each batch. The cocoons stay dormant indefinitely until the perfect conditions occur. Baby earthworms can only be about an inch or an inch and half long when they are born.

Sexual Maturity, Death

This could be described as the second half of an Earthworm's lifecycle. I have read that some earthworms can live up 8 years in natural settings. I am not sure how long they will live in a farming environment.

Several things occur as earthworms develop. The formation and growth of segments begins. These giant segmentedworms, which we are used to seeing, are already fully grown. Their rings and bright color are not present in their young years. This is why many people don't recognize that they're actually looking at an Earthworm.

The deeper an earthworm's rings become, the older they get. A fully grown adult earthworm will have 120- 170 segments. The clitellum formation process in an earthworm takes place between two and three weeks. That means they can breed in three to four week intervals.

After six weeks, a worm has reached full maturity and will stop growing. They are able to slide through tunnels and have many children. Depending on the species, worms can live up to four to eight more years. Lumbricus Terrestris is a species that can live as long as eight years.

Chapter 14: Types o Worms For Vermicomposting

Vermicompost is more rich in essential nutrients like potassium, nitrogen, and phosphorus than traditional backyard compost. Because of the worms, vermicompost also contains more beneficial microbes.

Vermicomposting is best done with common nightcrawlers, red wigglers, and other small worms. Consider that you require a worm with greater efficiency in decomposing organic matter or kitchen scraps. In such cases, red wigglers may be the better choice.

The United States Department of Agriculture (USDA), published an article. I learned that there were over seven-thousand types of earthworms. That's a lot! The lengths of the species vary between an inch and two meters. The decomposers of organic matter, dead and decaying, are earthworms.

You could argue that they're vegetarians since they don't eat meat and dairy.

Nightcrawlers

There are many varieties of nightcrawlers. The garden is well-served by grunt worms or garden worms. They are also a highly sought-after worm bait. Nightcrawlers are a good source of food and protein for many reptiles as well.

There are three kinds of nightcrawlers.

European Nightcrawlers (Eisenia hortensis)

Canadian Nightcrawlers (Lumbricus Terrestris)

African Nightcrawlers - Eudrilus eugeniae

People use them as fishing bait because of this. Even as children, the nightcrawler catch was a lucrative business opportunity for those in the local neighborhood.

All three nightcrawlers are beneficial for the soil, ecosystem, and wildlife. But it's important to remember that African nightcrawlers live in warmer environments to thrive outside. All three of these could be worthwhile investments if your primary focus is on the business of worm farm.

European Nightcrawlers (Eisenia hortensis)

Fun fact: European nightcrawlers have a cousin in the red wiggler. The main difference is the fact that a nightcrawler weighs more than its smaller cousin. A second difference is that the nightcrawler can live underwater for longer times than its smaller cousin.

They will burrow in most compost piles and in piles containing animal manure. European nightcrawlers are approximately 2 to 3 inches long.

Canadian Nightcrawlers (Lumbricus terrestris)

Canadian nightcrawlers average 5-8 inches in length, making them the second largest species of the three. Some call them dew worms. The Canadian nightcrawler is also known as a deep burrower. They prefer to position themselves in smaller spaces, but I suspect they are happy with their freedom.

African Nightcrawlers - Eudrilus eugeniae

Another tropical worm that is found in the area is the African nightcrawler, Eudrilus eugeniae. These worms can survive higher temperatures if they are able to retain enough moisture. With worm bins, this environment is easy to achieve. An African nightcrawler will not tolerate temperatures below 45°F. We will need to have dry, temperate conditions.

The Eudrilus Eugeniae has a length of between 8 and 10 inches. They normally burrow six to eight inches above the topsoil. A African nightcrawler can burrow a little deeper, if needed.

Blue Worms - Perionyx excavatus

If you live or work in a tropical environment, the blueworm may be an option to your worm farming operation.

The blueworm is also known under the names India blueworm and traveling blueworm. They can grow up as long as 1 1/4 to 2 3/4 inches. It is important to maintain temperatures between 70-80 degrees Fahrenheit. Blue worms reach maturity in about 3 to 5 week. You can positively affect their growth and productivity if your plants are in a warm environment or have a tropical garden. You could also sell them if the blueworm is not used or needed.

The jumping Worm (Amynthas.gracilis)

This wriggly crawler comes with a warning. Jumpingworms are an invasive species that is native to Asia. Jumpingworms have many names: snake worms; jumper worms in Alabama and Georgia; crazy worms in Georgia.

You can find out the history of a jumping bug by disturbing it or causing them to jump. These suckers will be violently shaking on the ground, acting as though they are experiencing something terrible. They'll rip off a portion of their tail when you try and catch them.

Jumping Worms are invading and can do more harm than good. Because they live on the topmost layer of soil, they eat the leaf litter or organic material. The top layer of organic material leaves little to no nutrients and is therefore not recommended for plant growth.

They leave behind very little soil and steal the trees' food. Gradually, erosion is a problem that will need to be managed. Jumping Worms can damage the roots of young plants. This is another problem to be aware of. If you are looking for worms to help your farm, be aware that a few can cause havoc.

Red Wigglers (Eisenia fetida)

The redwiggler is a worm species that can be used for vermicomposting. Red wigglers are able to eat half of their daily weight in food. They have been shown to be very efficient at breaking down decaying organic remains or material. They turn kitchen scraps and other organic matter into high-quality compost.

Because they can eat so much, they are able to produce castings fast. All of this is topped off by their prolific nature, making them a top-tier choice for your yard and worm box. Red wigglers are able to grow up five inches in height, with an average of 1/4 inch.

If you're new at worm farming you will need to determine why you are doing it before you can decide on the best species.

Chapter 15: The basics of Worm Bins

An assortment of materials can be used to make a worm-bin. Or you can buy one. There are four basic styles or types of worm composters. Each style is different.

Traditional Worm Bin: Bins come in any shape and size, but the most common is a box. These containers are used to raise worms indoors as well as outdoors. Wooden or plywood are two options for building bins. A popular option is to use old household items and empty totes. I've seen old fridges, empty water tanks, busted toilets, old tires, and other household items being recycled into worm farm containers.

Flow-Through - A flow-through is similar to a bin, in that food scraps are able to flow through it downwards. The bin can be used indoors or out. The main difference is that the worm casts are collected from the bottom of both bins. Castings in other bins are collected from the top layer.

Stackable Bins Each stackable container is usually equipped with a cover to protect it from pests or prevent any spoilage. Each tray serves an important purpose.

Beds: Outdoor bed worm bins require nothing but cinder block, brick, or wood. It basically means that you have set aside some earth to grow worms. While it is funny, the beds used as bins are free-ranging for worms. The best and most organic method.

Before you rush to order your worms, choose which bin or system you would prefer to work with. A majority of the time, beginning worm farmers will find something online or in a store that triggers a compulsive purchase.

I'm not saying it's a bad idea. Here's some info. You should be aware of two important things when researching condos or worm communities.

First, pick whether you want to use a commercially made bin or a handmade DIY bin.

Second, you need to decide where you'll keep your worm farms.

Once you have answers to these questions, you can then start looking at different farming methods. However, you can still order a container and start farming.

Traditional plastic worm bins

Traditional worm boxes are still the most widely used method for raising worms. The possibilities are endless when it comes to what you can put in a bin. However, rubbermaid plastic and Sterlite bags are the most widely used. You may have seen a few ideas for upcycling mentioned earlier. But here are some more. Buckets (or restaurant tubs), cat litter tray, and almost any plastic container that holds two to more gallons.

Worm bins usually have covers to prevent your worms getting out and keep nuisances out. If your worm container does not have vents, I strongly recommend making modifications. Drainage holes, spigots and drain holes are added to drain excess moisture and leachate. Easy harvesting is made easier by ventilation and drainage. Too much moisture could lead to mold. You'll read more on this later.

Traditional Plastic Worm Bins - The Advantages

They are also very economical.

Easy to create

Somewhat stackable

Excellent for indoor worm farms

For outdoor farming, it is barely allowed

The downsides of traditional plastic waste bins

They should be indoors when it is cold

They can be very heavy.

Hard to Harvest Worms and Castings

Plastic Worm Bins Are Easily Waterlogged

Plastic worm bins in small spaces are great. If they are ventilated properly, they will be less likely become waterlogged. Ventilation also addresses the issue of heavy bins that are difficult to transport. You need to ensure there are enough holes so rodents don't get through.

Traditional wooden worm bins

Perhaps you are looking for something that is more visually pleasing or you prefer to use natural materials. Traditional wooden worm bins might be worth your consideration. You can make one or purchase one online or at a garden centre.

Traditional wooden bins are nothing more than wooden boxes with lids, drainage systems and bottoms. If you decide to make wooden DIY bins, be sure to use eco-friendly products to waterproof them. Toxic materials can leach into your bedding, or your bin, and could cause harm to your worms.

Traditional Wooden Worm Bins Have Their Advantages

They are highly effective

Easy to use and set up

It can be made with scrap lumber

It's easy to move about (depending upon the size).

The downsides of traditional wooden worm bins

An insufficient amount of insulation

Slatted styles allow liquid to drain off and leave the compost dry.

Wood will decay if not safely treated

They can't be vermin proofed without modifications

Wooden wormbins made of wood are best for outdoor use unless there is a drip or drain plate underneath. You can make them any size you want. It is important to remember that wood without weatherproofing can lead to rot and mold. The lack of insulation makes it difficult for them to be kept warm.

Metal Worm bins

I will venture a wild guess that this metal bin was built out of necessity rather then preference. They're not necessarily traditional, but they are still used occasionally. I do not recommend that metal be used for bins.

Metal is a good conductor of both heat and cold. It is important to keep the bin's temperature at all times when you use metal.

Metal is also susceptible to rust. This is especially true if it is exposed to moist conditions such as worm farms. If you do decide to use a metal-worm bin, I recommend stainless steel or galvanized. Avoid tin, aluminum and other sheet material.

The only thing I like about a metal wormbin is its ability to be used in all situations. Other than that it's pretty straight forward. Metal bins can be more time-consuming, expensive, and require more maintenance.

Metal worm bins cost as low at $15 for a basic start, and as high as $100 for more complex projects. There are so many items you could recycle to make traditional bins if this is something you are interested in.

Flow-Through Worm Bin System

A flow-through, or continuous flow, worm bin system was designed to appeal to the natural instincts of red worms. The worms reside near the earth's surface so the food and worms are added at the top. The worm casts will naturally flow down to the bottom of the bin, where they can be harvested. Removable bottom tray may make harvesting much easier.

A typical flowthrough system is constructed from either wooden boxes or plastic containers. The bottoms are left exposed and a network wires, pipes or dowels is installed to cover them. Whatever material is used, it is placed horizontally along the bottom edge of the bin. To stop the bedding or worms from getting through the gaps, cardboard or thick paper is used as a top layer.

Pros of Flow through Worm Bins

They're extremely efficient

Easy to harvest the casts

Easy to maintain

The indoor closest system to the earthworm's natural habitat, is a continuous-flow bin

Cons of Continuous-Flow Bins

A prefabricated system will require a large initial investment.

They lack mobility. They are generally large, bulky, and often pack some weight.

Many advantages can be gained from continuous flow-through system. These systems are usually made of wood but can be made out plastic. Rubber or plastic flow-through containers look just like trash cans, but they are much easier to transport. The cost of this system for residential use is anywhere from $120 to $400, and for commercial use it can easily reach thousands.

Stacked Bins

Stackableworm bins save space. The stackable worm bins are made up of a series or trays that are vertically stacked on top. A stack of bins rests securely on a base unit, which doubles up as the moisture collector.

You'll see that stackable containers have many holes in the bottom. The holes allow the insects to migrate to the next bin at the right time. Once the worms' food and bedding have been transformed into castings the worm instinctively moves to a higher place.

I should also mention that most worms will crawl from the bottom. This is because they are always looking for the one working backwards. You can harvest them easily because they travel up rather than in a traditional container. To harvest your bounty, all you need to do is take out the old tray, clean it, and then replace it again with a new one.

The moisture retention is great when the trays are stacked on top one another with a tight fitting lid. The base of stacked worm containers usually has a spigot, drain or other device to drain off the leachate. This allows you to collect your worm tea.

A decent stacked-worm bin system can be purchased for between $50 and $100. You can find plans online or build your own. It will be less expensive and you'll feel more accomplished. The most basic DIY materials are wood, plastic, and cat litter bins.

Worm Trays

I am aware that worm trays are not on the above four-part list. This is because they aren't actually a worm container. These trays look much like worm bins. Worm trays work differently from traditional bins.

Worm farmers using worm trays are most probably breeding worms. These worm trays can be used by large worm farms. While castings are made by worm trays and casts can be obtained, reproduction in a controlled environment and following a strict diet is the key focus.

The worms are only able to spend enough time in their genes to produce cocoons. They then get harvested and relocated to another set, where they will continue the cycle. The cocoons in the bedding are removed until they hatch. After that, a new tray is made with worms to begin their lives.

Outdoor Worm Bin Systems

Similar to indoor bins, outdoor bins also come in different styles. There are two major types of outdoor bins, buried and above-ground. Windrows, worm beds and other methods can be used to grow outdoors.

Above Ground Outdoor Worm Bins

The Big Worm Rotter is one brand of above ground or on-ground garbage bins, which are often used outdoors. You can make your own, or give it a cool name.

Above-ground outdoor compost bins are often made using large containers or a flow-through system that has been insulated. It is important to control the temperature of outdoor bins. To keep it warm enough, many people use a temp control device.

In order to have outdoor bins, insulation is essential. In colder climates, your worm farms could easily become uninhabitable.

Above Ground Outdoor Systems: Pros and Cons

They will attract worms to their area and feed them

You can even use the garden as bedding

Easy maintenance and easy upkeep

You don't have to feed your pets as often

If their bedding or food is not safe, they will return to the soil and retreat until all is well.

Harvesting is not necessary. You can simply move the bin and get started in the next crop.

Cons of Above Ground Outdoor Systems

Climate control and temperature control are critical.

Outdoor predators or pests are more likely that they will try to infiltrate a bin outdoors.

For bins to stop rotting, exterior waterproofing and weatherproofing are essential.

Systems of In-Ground Bins and Buried Bins

You can find in-ground worm boxes in windrows or buried bins. This type of bins allows the whole system of worms to be contained in the ground, along with their bedding and food.

Buried Bins

A buried container speaks for itself. This is a worm bin that is flush with the ground and used as a contained system. The worms will not escape as long they are kept in a safe environment.

Windrows

Sometimes windrows may be called very trenches. These are straight, long trenches dug into the earth. After the dirt is removed, the trenches will be filled with food and bedding. The trench can be filled with decomposing vegetable matter, shredded cardboard, and horse manure. A second layer of bedding can be added after the worms are placed in the trench.

Winter months are when worm farmers insulate their windrows with leaves, straw, and/or both. They use the insulation to cover the whole top layer of the farm. It depends on where you live, extra insulation might be needed if you have to deal with winters that are particularly severe. Insulation is essential for worms' survival.

The advantages and disadvantages of in-ground systems are much the same as those above-ground. Mother Nature can be unpredictable and many parts could cause us to fail. Outdoor worm bins mean you don't have to create a space inside your home, and it's easy to clean up any mess.

You can store more worms in larger bins. But, you will need to be able and able to move them around on your own. It doesn't really matter whether you're growing indoors or outside, the size and shape of the bins you use is your choice. You can make them as small as necessary.

Chapter 16: How to Build a Worm bin

I have included several DIY vermicomposter or worm composting container instructions. Depending on what materials you use, a worm garden can be very affordable to construct. Anything made out lumber will cost you more unless you have it. The more complicated the bin's design and structure, the more you will spend.

The Basic Plastic Worm Bin

Plastic worm containers are best used indoors. While they can still be used outside, they should be kept indoors. The summer heat can cause them to become too hot, and worms may freeze during winter.

It is best to keep an indoor bin out of reach. A spillage could cause serious damage, so make sure you keep the container out of reach. Even if you are making worm tea, there will always be mess. Unless you move the bins outside to harvest castings or tea. As long as they don't get too hot/cold, you could keep them in your garage or basement. A second option is to put the bins on a tarp or in a larger container.

Supplies & Materials List

Keep in mind that you may already have these items around your home!

Two plastic bins: Each one should not be shorter than the other. It is important that the taller bin can fit within the shorter one. The bottom tub should be approximately 15 inches deep, 25 inch wide, and 5 inch high. A 15-inch deep, 15-inch tall, 15-inch tall and 20 inch tall 18-gallon tub works great for the actual hotel.

Screen Material, Mesh Screen: This could be used as a screen for your windows. Important is to not use any material that can rust when exposed over time to moisture. It will work well with scraps of 4-inch by four-inch screen. This screen will keep your worms safe from getting out for the night.

A drill: You'll need to have a drill and drill bits that are 1-inch in diameter and 1/8 of an inches in diameter, respectively, to drill ventilation and aeration hole.

Waterproof glue is used to secure the screen even after it gets wet. It is important to use non-toxic glue, which is both safe for the environment and safe for your worms.

A trowel or garden spade: This is only to move the soil and compost around inside the bin.

Shredded Newspaper Once a week, you will need to feed the worms more paper. It doesn't matter what kind of paper you use, but it's better to avoid using any color or dyes. You should also avoid anything too thick, heavy or shiny.

A Bit of soil: You will need about one pound of dirt. The worms won't wait long to start creating their own compost soil! You should ensure you are using only clean soil that is free from any sharp or harmful materials.

Water: Only enough water is required to moisten dirt and paper. Your worms will thrive in this environment. It's important to soak your paper in water, then drain it.

Worms - A pond of live worms is sufficient to begin with. Many recommend that you have a pound red wigglers, as they eat more waste than earthworms. I think half a pound each is a good choice. These can be ordered online at the USDA or other companies that worm farm. Check your local bait shops and pet shops to check if you can find them.

Food Scraps: Gather fresh produce scraps to feed your worms. Be aware of what foods are best avoided when saving scraps.

A Note About Bins & Holes

The holes near the top allow air to flow in. The holes at the bottom of the tub allow liquids and worm teas to drain. Your worms can drown if they drink too much. Although you could manually manage excess liquid, it can be more difficult. So, the holes are the best solution. It would be horrible to lose the entire farm if you forget to drain it.

Screen material will be used to cover the top and bottom holes. This will ensure that your worms remain safe. The lid is not required for the shorter bin, but the taller one will. If you don't have an spigot or tap, the shorter bin at the bottom will allow you to scoop out your worm soup. A bin should be available for your hotel bin. Because it is easier to drill holes, the lid should be flexible.

Prepare your bins

The instructions given below are simple and should be easy for you to follow.

On the one side with the taller bin drill a hole of one inch about 2 inches from its top. On the other side, drill another one-inch hole. Turn your container over and drill four 1-inch holes near the bottom.

You can cover all the holes with the vinyl screen and then glue them in with the safe, non-toxic waterproof glue. Allow the glue drying completely before proceeding to the next stage.

Place your tall bin into the smaller one. No matter what you do, don't drill any holes in your smaller container. This will cause a lot of mess.

It's easy. Some people prefer to add a fixture to the bottom of the bin to make the worm-tea extraction process easier. If you do decide to go this route, make sure to place the worm tea container on some bricks. They must be high enough for the runoff to pass into whatever container they are going to go in.

Final Touches & Preparations

It's now time to make sure your worm has a bed and breakfast. Be sure to soak the shredded papers in water, then drain them. Then, combine the shredded papers and soil. Only add enough water to moisten the entire mixture. You can add water to your shredded papers if they are still wet from the soak and the mixture feels damp.

Then, pour the mixture into a container and fill it with about three inches of water. Once you have the container filled with your preferred bedding, put your worms into their new home. You don't want them to eat right away. It is best to give them time to adapt to their new surroundings before you feed them. The bedding should be moist enough to not cause puddles, but not so much that it becomes a slurry.

Maintaining the Plastic Bins

If you don't have an outlet for draining the tea, you'll have to take it out once a month or as it accumulates. The following steps should be taken once your worm box is completed.

You can feed your worms one side of the bin for several weeks. This will allow the worms move to the preferred side.

After the worms are gone, you may harvest the compost by going to the opposite side. The compost should be collected at the end each week so that they can be fed again.

If you feel like there are too much worms within your bin, add them to your yard, share with friends, and/or start another one!

Bin maintenance can look a lot alike, but there are slight differences.

DIY Stacked Worm Bin Systems

You can always purchase a stacked system of worms from a reputable worm farm. Prices can vary depending on the size of the seller. The cheapest I've seen was around $30. Prices can rise into the hundreds. If you can make your own, why pay twice the price? You don't have to be very handy and you don't even need expensive materials. You can still design a well-designed, functional unit.

The stackable bin systems are ideal for spaces with limited space. They're also easy to use. The idea behind a stackable multi-tiered system is to connect worm trays or bins. The trays must be tapered slightly so the bins can rest in one another. If the bottom tray is getting too full, empty it and turn it over.

How to Choose the Best Nesting Bins

There are many sizes, and even numbers, of nesting tray you could choose from. Your desired scale of operation will dictate the size of the nesting tray you choose. Because they are available in different heights and widths, storage bins work well for this worm system. You will need a lid to cover the stack. Worms do not like light so it is best to avoid opaque bins. Stick with dark colors. Black works extremely well!

Three ten gallon totes are the best choice for small-scale setups. This size is perfect for composting small amounts. Larger bins have more tiers and are better for composting large amounts of kitchen scraps. You could also choose smaller totes and buy more. They should be mobile enough for you to carry them around.

Supplies & Materials

At least three bins

One Lid for the top tier

Drill with a 1/4 inch drill bit

Mesh or Mosquito Netting

3/8" Drain Spigot, or Barrel Tap

Wooden Spacers 6- 8 Inches Long

Instructions

It is not difficult to construct a stackedworm farm. It will take some extra work but it's much easier to manage and maintain.

The Sump Bin

The sump bin is located in the bottom of a stacked Worm Farm. It is slightly different than the upper bins in a stacked worm farm and should always be prepared first. The sump container is used to collect the worm soup.

The barrel tap or 3/8 inch barrel tap may be installed on the sump tray. To install the barrel faucet or spigot, a small hole will be needed near the base. It doesn't matter if you have a fixture. However, it will be easier to put in. If you do decide to use a tap/spigot, ensure it is well sealed by adding washers or locknuts.

The Worm-Composting Bins Preparation

Your worm's habitat will be the two top or higher bins. Here are the steps to create them.

Each bin should have 1/4 inch holes to allow for ventilation. Your holes should be about two inches apart and centered in either direction.

Now drill a row at 1/4 inch intervals in a straight line along the walls of your composting bins. The holes should be at least 4 inches below each worm bin's top edge.

This is all you need to do for this step.

The Set-up

After you've got your composting containers set up, you can start to put them in place. It is important to choose a place that will allow your worms the best chance of survival.

You can make a base for your worm farm using bricks or blocks.

Place the sump on top. Be sure to leave enough space between your sump and the ground. It is important that the worm tea can be harvested from any container.

Now that your sump is in place, you can take the two lower compartments and nest one inside another. Then, put them in the sump. For the space between the two upper bins, you will need a few spacers that are 6 to 8 inches high. This allows for compost accumulation and provides space for the worms' movement. As long as the job is done safely and effectively, you can decide what kind of spacer to use. The spacers also prevent bins from getting jammed together.

You can cover the gap between bins using strips of shade cloth and mosquito netting. This will prevent pests and insects from invading your worm farm. It also helps to prevent worms moving out.

Wooden Worm Bin

Wooden worm containers are much more attractive than plastic, and they can be weatherproofed better than plastic. This type of bin can also be used outdoors. I enjoy the natural look and feel of these bins.

Because it is naturally weather resistant, untreated cedar repels most insects. It is also possible to use scrap wood or exterior siding. Any wood that has been treated with chemicals should be avoided as they could leak into the worm farm and make the environment toxic. I'm certain you will find a section with safe methods to weatherproof, waterproof, insulate and insulate the farms.

Supplies & Materials

This list will go a lot deeper on the materials. This is mainly to let you know what you're getting, and its purpose. You never know what you can do to make things better.

12 - 4x4s with a length of 7 1/4 inches (for blocking the bin corners)

8 - 1x4s of 11-inches length. These will be your bottom bin legs.

16 x 1x4s, 7 inches long (for the top bin legs).

2 - 1x4s 12inches long (for the inner crosses-pieces for lid)

3 - 1-inch wide, 21 3/4-inches in length (for the outer cross-pieces to cover the lid)

12 - 1x8s, 48 inches long (for the floor and lid)

6 - 1x8s 20 1/4 inches in length (for short bin sidewalls).

Assembled with nails or screws

A pair small hinges and hardware

Drawer handles or knobs for hardware

2 pieces of wire or hardware cloth 48x22 inches (for the bottom of the upper bins). 4x4 hardwood fabric, 4 inches per square works well

Hardware cloth can be attached using screws and large fender washers

2 hinges, and a handle

One container for the worm-tea (3 aluminum baking sheets will also work)

A Jigsaw

Instructions

Let's start these bins.

If your lumber has not been cut to the desired sizes, you can do so now.

You can build the three-sided, 4-sided bins.

(2) 48-inch sidewalls

(2) 20.25-inch sidewalls

The 48-inch sideswalls are to be exposed at their ends. The shorter wall should be sandwiched at its ends. To nail into the corners, use a 4x4 blocking that is standing straight up. It is important that each of the three boxes you use is exactly the same size. You can verify this by stacking them together.

To add a floor on the bottom bin, use three 48-inch boards. Nail the boards to your frame. Nail one edge along the long side. This space should be between the 4x4 blocking piece. This allows for you to create an access panel that will allow you to use drip pans.

At the length and height you prefer, mark the location where the panel will be placed. To carefully cut the panel, use a jigsaw.

The hinges can be used to attach the panel door back to the sidewall. To allow the panel to open or close smoothly, it may need trimming or sanding. If you have extra space, you could add additional cross-pieces inside the panel to hold the floorboards.

Attach the bottom bin legs by using eight of your 11-inch boards. Two boards should be placed on each corner. You should place your panels so that there is at least 4 inches gap between the tops of your bins and the tops of the legs. This step is critical because it allows the upper containers without any space to be stacked on them.

The legs will attach to the upper bins the same way they did the lower. For the upper bin, you will need to use the 16, 7 inch boards. You should leave the same spacing of 4 inches as for the bottom bin. You don't need to make the gap large. It is not essential that the legs from one container slip over the legs of the next, so they stack tightly without interference. Your legs may need to be adjusted by repositioning the legs as you fasten them, or trimming if necessary. You should make sure that the bins are properly aligned before making any adjustments.

Your lid will be made by assembling your three 48-inch boards. You can attach the boards using nails or screwdrivers through three of the 21 3/8 inch cross-pieces located in the center and at the ends. Fasten the two 12-inch sections to the underside. You should ensure that your cross-pieces do not block the corners of the 4x4 blocks and the sidewalls. To make it easy to remove the lid and place it back, you can attach handles.

Now, attach your hardware cloth and netting to both the bottom and top sides of the upper boxes. If needed, trim excess mesh. Fix the mesh to frame and 4x4 blocking with screws through grid squares. Use matching fender washers that are larger than squares to fasten the mesh. The top bins must look identical.

You're done! Now, all that's left is to place your worms in the bins. Keep in mind that a wooden farm should be outdoors so you can waterproof it.

Bins with In-Ground Buckets

This might be the right project for you if your interests are simple. This is a simple project that can be done quickly. This is an indoor worm bin, which will be underground in your garden. You can use the in-ground bin or bucket method to prepare the soil if your garden is not yet ready.

In this example, the ground acts to provide insulation for the worms. It's important that you remember to check your trash bins every day.

Supplies & Materials

The only thing you will need is the bedding, compostable and worms. If you have the space, you can build multiple buckets.

5 Gallon Bucket with Secure Fitting Lid

Drill and 1/3 inch Drill, or Spade Bit

A Shovel

Instructions

If you already have your own garden, select a spot that is suitable for your worm-bucket. Call 811 to verify that there aren't any hazards in the area, such as buried electric wires or other potential dangers before you dig.

Once you have chosen the best spot, dig enough holes to allow the bucket to fit into. The top should sit slightly above the soil.

Make 20 holes in the bucket using your drill. Then drill 10 holes in the bottom. DO NOT drill holes through the lid.

Bedding of your choice should be aired. Use damp leaf litter or soaked shredded paper. To provide worms with the necessary grit for digestion, mix in some soil or sand.

Now you can add your worms. It is common to hear people refer to adding compostable food wastes. However, it is better to give your worms 24 hours to adjust to their new environment before feeding them.

Place the lid securely on the worm farms and place the bucket inside the hole. To ensure the hole is tight and secure, add some dirt around the holes.

After 24 hours, it is time to give your worms their first food in their new home.

Keep in mind that the inground worm-bucket method can be left to run for a while longer than most above-ground systems. It's still important to monitor it. It is not something you can just "set it to forget it". If left to its own devices, the castings may become hardened and almost fudgy. It is possible to empty the bucket straight into the hole, and then start a new bin. You'll love your garden for it.

It is recommended that you stop adding food to your worm farm around 2 to 4 week before harvest. But don't worry! Your worms can still enjoy plenty of food and enough time to finish the meal before harvest. When your worms become free to roam in the garden, there should be plenty for them to eat. The bucket can be placed on growing trees or in the same place you took it out of.

A Few Side Notes

I did mention earlier that there are many species of worms that can be used for vermicomposting. If you are using the inground bucket system, it is best to stick to the redwigglers. Nightcrawlers may not work, but redwigglers are voracious eaters. They can transform your garden in a matter of hours, compared to other types of worms. They are more likely to be at home and not roam around the garden. This isn't a bad thing. But they do have a job.

Your worm farm should be checked for moisture every once in a while. It's important that the contents feel as moist as possible. If the contents are dripping, they're too wet. It's best to add some water if the contents are too dry.

Finally, give your worms some time to adjust in their new home before you feed them or disturb them. It is important that they adjust to their new environment naturally. They should be allowed to rest for at most 24 hours. But, they can take a few days.

In-Bed Worm Composting Technique

In-bed vermicomposting is the act of composting directly into your garden. This farming method makes vermicomposting easier. Your worms are able to live, work and reproduce in your garden. The in-bed method of worm farming is basically free-ranging your larvae right where it's needed most.

You can farm worms directly from your garden. This eliminates extreme temperature problems. If and when necessary, your worms are able to dig deeper.

In-bed composting is very similar to the inground method. We would use a bucket. This method works best when used with raised gardening beds. If temperatures are extreme, you should not harvest castings nor move the bins.

Supplies & Materials

2-gallon Buckets with Fitting Lips

Shredded cardboard

Your Kitchen Scraps

Worms-Red Wigglers or Worms Do well

A Drill with a Large Bit

A Dremel

A Shovel

Instructions

You will need to drill several holes around your buckets. Drill all the holes around the sides of the buckets. However, don't drill in the rim.

Grab your Dremel and take the bottom off your buckets.

Dig the holes in the raised soil so that the bucket fits snugly, but ensure the lid and top don't go underground.

Put your bucket into the holes. Fill them with soaked, shredded cardboard. To get the best results, allow the cardboard to soak up the moisture over night.

Then, in the morning, add 300-600 worms.

If you have worms that were not already in the soil, it is a good idea to mix some dirt or sand with them before you add them to the bucket. A larger bucket or several of the two-gallon buckets may be required depending on how large your raised garden beds are.

You can always be a rebel by saying, Forget the buckets. It is possible to let your worms roam free in your garden, and not worry about barriers. Although they may roam freely, worms will be more likely to stay close to the garden beds if they know where they can find food and feel comfortable. You can always give them more worms if their water runs out.

You can see that the core components matter, regardless of what type of worm box you use. It is vital that the core components of worm bins are well ventilated, provide food scraps for the animals, and allow them to live, grow, and breed. Coffee canisters are an option. Some people even place a big cardboard box directly into the ground. Although the cardboard boxes will eventually decay and release the insects into the ground, that was not the purpose of the whole thing. To increase the worm populations for the gardens.

Adding Worms to Food

Some worm farmers suggest that you give your new worms some time to adjust before you start feeding them. The theory is that the worms can adjust to their new environment by waiting. Some farmers feel it's okay to feed them right away after they arrive at their new home.

Either way, I believe it can work just fine. Overfeeding your new worms immediately is one of the biggest problems. While it may be tempting to toss all your food scraps into the trash, it is not a good idea. Worms that are fed too many food scraps can become rotten, leading to an unhealthy environment and a bad smell.

I suggest that you wait at least a day if it makes you feel good. Consider if you feel it's necessary to feed your child immediately. In this case, I suggest tossing only a few items. It would be annoying to discover your whole NEW bin is full of trash. Your efforts would be wasted and your environment could become unhealthy for your worms.

Soon, you will find out more about feeding your worms. What to feed them, what not to do, and how frequently to give them scraps. While vermicomposting may seem simple, it can become difficult if you aren't careful.

Chapter 17: Winterizing & Insulation

The environment is the best place for worms to grow up. If it becomes too cold, they dig deeper into soil to get warmer. It's a different story when they are born and raised in captivity. A worm can only go so deep. This means that they can't dig down as deep as necessary if it is freezing outside. However, we must be responsible for our worms and care for them at all levels. If you want your worm farms to be outdoors, make sure it is prepared for the elements.

I will show you how to make sure your worm population is prepared for winter outdoors. It would be awful to discover all your worms had died due to freezing. If this were to happen, your bins might need to restart from scratch. Let's make sure that this doesn't happen.

Will my Worms Survive in the Winter Outdoors?

Your home is really the best place for your winter farm. Unfortunately, that's not always possible. This is because there are many variables. The temperature in your local area will also impact the results. It will affect the quality of your bins, regardless of whether your bin is located inside a shed/garage or outside.

The best way to tell if it's too cold is to think about the temperature you are comfortable with. Similar to humans and worms, they prefer temperatures between 60 and 80 degrees F. It's best to stand directly in front of your worm bins. They'll likely feel the same way if you're too warm or too cold.

Now, don't get your hopes up. Although they are able to thrive in these temperatures, it doesn't necessarily mean that they won't be able to survive in lower or higher temperatures. Your worm colony should be active at 60 to 80 degrees.

Your worms can continue processing the compost at a slower rate if temperatures drop below 60 degrees Fahrenheit.

The most important thing to know is that worms cease to work when temperatures drop below 40°F. The goal here is to keep your compost bins above 40 degrees Fahrenheit all winter. I recommend getting a worm-composting thermometer to get readings from all areas of the bin. This will ensure that temperatures are never dangerously low.

Insulating Your Wormfarm

There are many different ways to insulate your worms. It is easiest to bring them inside. It is possible to keep them indoors if you have a garage or shed. There are some things you can do to make sure your worm bins stay outside.

Add an Insulating Layer to Your Bedding

As insulation, a layer of presoaked, shredded newspapers works extremely well. Before you add the newspaper, soak it. To soak up excess moisture, place a few layers on top of the moistened newspaper. The thermal barrier created when the paper is wet can reduce heat loss from the bins. Burlap bags and coconut-coir mats are also good options. It is important to add insulation before the temperature drops too low. Imagine how much it will make a difference to wrap yourself up in a warm blanket while you are still shivering.

Sidewall insulation

There are a few ways that you can insulate the sides to your worm farms. There are many options. You can dig a hole that is large enough for the bins to fit in, but don't bury them. It does not need to be very deep to accommodate the bins. Bales of Hay are another option. Just place the bales along the perimeter of the worm boxes and you are done. Finally, you can make an exterior box from blue board insulation foam. The foam board should be close enough that it can hold the heat, but not so close as to block the ventilation holes.

Heat will naturally be created when compost is placed inside the bin. It's important to hold the heat in place so it doesn't escape. Add some leaves to the containers. They will also release heat as their decomposition occurs. Your worm farm should be monitored regularly, especially if the temperature drops below 60 degrees. If none of those suggestions appeal to you, warm blankets might be a better option. They are quite effective, according to me.

Cooling & Shade

If it is too hot out, your worms will be too. If you live in an area where there is dry heat, make sure your worm farms are hydrated regularly. To be safe, wild worms will follow their instincts and burrow deeper if it gets too hot.

They don't sweat like humans or pant like dogs. Instead, worms get out of the heat sun, dry sultry atmosphere, and hot surfaces. The depth of worms' burrowing ability is limited when they live in a container or worm farm. It is our responsibility to keep them cool and safe. Their bedding must be fluffy and easy to breathe. It can be very difficult to breathe if it is too hot, dry or humid. Imagine being trapped inside a bin without adequate ventilation. I find it amusing that the same technology that cools the worm bins in summer can also insulate them in winter.

Simple Designs for Cooling or Bin Features

Vermicomposting bins should have all the features necessary to keep the worms happy. While these features should be part your vermicomposting containers, you can see how important they really are.

The lid should be kept on the bin. It protects the worms and provides the necessary moisture.

These holes regulate the farm's moisture and are critical. If your bins are too wet, excess water will drain out, keeping your worms safe from drowning.

Ventilation holes are designed to prevent too much heat from building up in the bin. Ventilation holes are important to keep the worms cool in your worm farms.

A warm blanket can work in both the summer and winter. If you don't already have a blanket, you may use soaked newspaper as a substitute. This technique will help preserve essential moisture.

You can place the bin in a sunny spot during the summer. In hot weather, your wormbin's temperature will increase dramatically if it is in direct sun. There are many options. You can place your worm bin in shade, such as under a tree, a porch awning or in a shed that has good airflow.

Important: Moisture

We add moisture to the worm bins by placing food scraps in them. The problem is that the bins may become too dry in the summer heat due to evaporation. Make sure to check your worm farms regularly to ensure they are moist. If it isn't damp enough to wring out a sponge, it may be too dry and you will need more water.

I would recommend that you check your bins at the very least once per day in hot weather. Don't overwater, as too much can drown your wigglyworms. You can add water, let it soak in and check the moisture level once more. You can add more water to make it moist.

You can always add a few more ice cubes in the bin. Some can be placed on top of the bedding, or buried in the middle. Another option is freezing the produce in a small amount of water. Place the frozen chunk in a bin or bury.

If your worms are too hot or left in direct sun, they will dehydrate and eventually die. There will be dead worms in the bin. Don't wait till it gets too hot to do something. If you take care to your worms before the game begins, they will thrive.

Waterproofing & Protecting Wooden Bins

You must use non-toxic materials for your wormbins. Cedar lumber is an excellent choice for building a wooden worm bin. It is both weather-resistant and waterproof. I am not a fan chemical solutions so it is natural for me to seek out all-natural alternatives. I've compiled a list of natural options for sealing and protecting wooden wormbins.

Olive Oil

Olive oil is wonderful stuff. For waterproofing wooden worm bins you can make a mix of olive oil and lemon juice. You shouldn't feed your worms lemons, but it doesn't stop you from using them for the exterior. Spray bottles are possible but I prefer to dip a cloth in the solution, then rub the mixture into the wood. You can rub the olive oils into the wood grain by moving it around. Let the wood absorb the oil solution for about ten to fifteen minutes. After that, use a clean cloth to wipe away any remaining oil.

Olive oil is a great natural option to seal your wormbins. If the wood begins to look duller, it is likely that you need to add more oil. If you do the job right, the oil will last quite a while. Only you will need to keep it in good condition.

Coconut Oil

Coconut oil can't be used as a water-repellant. However, it does have some benefits. Coconut oil can also be used as a high-quality wood conditioner. This oil is well known for keeping wood soft and protecting it against warping. The solid version and the liquid versions are both available. I prefer the solid because it has more power.

Mix the oil together well. You can use a cloth sprayed with oil and rub it into your wood bins. After you've covered it, let it dry for about 2 to 3 minutes. Next, wipe it clean with a dry, clean cloth. Coconut oil can be used to protect against elemental damage in each season.

Vinegar & Oil

It can be used to make many wonderful salad dressings. However, it can also be used to seal wood. This wonderful combination works well to repel burrowing insect invasions at your worm farm.

Take three tablespoons of canola oil, one teaspoon vinegar and mix them together. It's as easy as following the steps for other oils. The wood can be revived by reapplying the oil.

Linseed Oil

It is made of flax seed oil for those who haven't heard of it. It has a pungent smell and takes longer for it to dry. But it works well. You must ensure that the linseed oil is free of petroleum additives. Linseed oils are often used as a therapeutic food. But, they can also be used as an excellent natural sealant when boiled. You can either purchase linseed olive oil in raw form and boil it yourself, or buy it ready-to-use.

You will rub the oil into the wooden bins the same as with natural oils. However you'll need more time for it to dry. It's easy to forget about.

Protect your worm bins before they are damaged by the elements. If you are building your own, make sure you waterproof them before you add your worm farms. Make sure you check the details before purchasing a bin. Most worm farm companies will not use toxic or harmful products. However, you never really know. It's always better to be safe rather than sorry.

Chapter 18: Feeding Your Worms

Worms can have similar food preferences to us, it is true. People often believe that worms will eat whatever you give to them. This is false. It is common for people to believe that worms will eat whatever you give them. It's important that you know what foods your worms love, what's healthy for them, as well as what can be harmful to your goals.

Be mindful of how often you add food into their bins. Overfeeding worms can cause their food to rot, making them messy and smelly. If you underfeed them, your worms will be hungry and looking for food. After week two, your worms may start to disappear. If they don't get enough food, they might try to escape the trash bin.

As with us, our food is digested by bacteria. The ability to break down organic material and produce worm-sized foods is possible thanks to bacteria, fungi, and other microorganisms.

Best Wormfood Options

If you look at the list below, you will see that many of the foods recommended for worms have soft, moist food and low acidity. You should feed your worms a balanced amount of both brown and green food. Brown food has high levels of carbon and carbohydrates. Greens provide a lot more protein and nitrogen for the soil.

"Greens" refers to a combination of green fruits, vegetables, and other natural food sources. These items do not need to be green. However, the term "green," refers to freshness. You have many choices, such as fruit scraps (melon rinds), carrots and carrots. "Brown foods" can be food items or non-food items. These include paper, coffee grounds (egg cartons), dry leaves or grass, and paper. You can either grind the food in an electric mixer or manually. It makes the food easier to consume and also helps it break down. The longer it takes for the food to decompose, the more large they are.

What should you include in your worm's meal? This is the short answer to your question. But don't worry!

Fruit peels, and fruit scraps (NOCITRUS).

Melon rinds

Carrots

Coffee grounds

Biodegradable Tea Bags (make sure there is no staple, it can cut them).

Bread

Pasta (make sure it is plain).

Squash

Cornmeal

Lettuce

Spinach

Cereal (unsweetened).

Cucumbers

Eggshells

Grains

Worst Foods For Your Worms

Certain foods should be avoided or limited when you feed your worms. Like the foods they like, there is a list of foods that worms will not eat. Tossing meat, fats and grease in the worm bin is a good idea. These foods aren't good for worms. In fact, they can attract insects and critters to your garden.

Meat

Potatoes (including the peels)

Citrus fruits and their skin

Onions and skins

Animal feces

Fats – Grease – Oil

Fatty Foods

Dairy and dairy product

Salt

Processed Food

Sugars/Artificial Sweetener

Although worms can eat tomatoes, it's possible to find tomato plants growing after a while. You can either transplant them or pull them out and put them in the compost. Too many tomatoes in a bin can lead to an acidic environment. Two things you don't want are meat and dairy. They can become rancid long before the decomposition process completes.

How to feed your Worms

It's a smart idea to chop large chunks of food into smaller pieces before you serve them. A food processor or blender can also be used to reduce the food. This is a good idea as it speeds up the process of food decomposition and makes it easier on the worms to consume.

You can feed your worms every day or once a fortnight depending on how big the bin is. The larger your worm bin, the more they should be fed. The amount of food you need will depend on the size of your bin and how many worms you have.

You should place your worms' food approximately 2 to 3-inches below the surface. This will prevent odor and fly problems. Take out some of your bedding and put a layer of food over it. Finally, cover it with the bedding. Refresh the bedding every time you feed. In approximately a month you can add more bed.

It is a good idea to keep a journal in order to keep track of things like how quickly your worms are eating certain foods. It can help you adjust how frequently or how little food you feed and which types of food you prefer to feed your worms.

If your worm bin smells, it could be a sign that they are being overfed. Rotate your bin's feeding areas to ensure all your worms are getting enough food. To discourage flies, it is a good idea to tuck food into their bedding for three to four inches.

If you notice that the number and size of the worms is increasing, you can be sure you're properly feeding them.

Remember to feed your worms. One, it's unacceptable to leave your worms hungry and uncared-for. They will begin to seek out food and escape after a few more weeks. You should check your worm bins at least once a day to ensure they are fed the right food. If they're not eating it, take it out of their bin and toss it into the compost pile. Over time you will discover what they want and do not want.

Chapter 19: Worm Bedding 101

One of the most important components for a healthy farm is worm bedding. For worms to be happy, healthy and productive, bedding must be added to their worm boxes on a regular basis. Inadequate bedding can lead to a host of problems, including death. Effective bedding is vital for the health of your worms.

What kind can you use for worm bedding?

Many types of bedding are available. It is important to use a combination or a variety of materials in order to make your system function efficiently. This is because different materials are better than other. By adding multiple types to the bins, you can overcome any weaknesses that might be present with certain materials.

Bedding Materials Worms Will Love and Admire

Maintaining a healthy worm-bin environment requires that you use both the highest quality bedding as well as a variety and frequency of materials. Many vermicomposters could learn from this bit of advice.

You can achieve the best results by using a combination of the following materials. Keep in mind that different materials have strengths and weaknesses. However, it is good to have the ability to use whatever materials are available. Your worms will eventually thrive when the conditions are right for them. You will be a good example to your worms if you are attentive and aware of all these things.

Paper & Cardboard

As bedding materials, shredded newspaper or cardboard make excellent choices. They are not very nutritional for the worms. However, they can be a great source in carbon.

To speed up the process of paper and cardboard decomposition, they should be less than 2 inches. I have found that soaking the materials in water before shredding reduces paper dust. Avoid glossy or highly colored paper. This can pose a risk to your worms. While white office papers can be used, the environment takes longer to process.

Straw/Leaves/Leaf Mold

You want leaves that can be broken down quickly. Poplar, box elder and maple are good options. Tannins can be harmful in high concentrations from oak leaves. You should not use oak leaves in your worm bin until they have had time to compost.

Leaf mold, leaves and spent straw are all possible ways to introduce other organisms to the bin's eco-system. Some pests may be a problem. Pine needles contain harmful acids which can be dangerous to the worms. Also, they take a very long time to break down. For faster decomposition, you should shred or chop as finely as possible.

Coconut Coir Fiber

Coconut coir is made from the hulls, or coconuts. Coconut coir fibre is most often recommended for use as a bedding material. It's neutral in pH, and holds 8 to 10 times as much water. Coconut fiber can be used as a renewable substitute to peat. You can mix it for potting mixes and amending your garden soil.

Peat Moss

Peat moss costs very little and can be found at many garden centres. Peat moss can be very acidic and should be adjusted pH with ash or lime. Peat is neutralized by adding ash and lime. Before you add peatmoss to your compost bin, ensure that it has been amended correctly. You must inoculate your peat with compost or compost tea before you can use it.

Green Manures: Grasses/Clovers/Plant Material

A nutritional boost is provided by green manure made from grass clippings and wild herbs. Sometimes, green manure can also be called "green undecomposed material". These bedding materials create heat when it is used fresh or in very high quantities. Green manure can also be dried and shredded prior to use. Be sure to use this bedding in moderation.

Manure & Animal Waste

Only use manure from herbivores. The best choices are horses, goats, alpacas, rabbits, goats, and llama manure. Llama, goat, alpaca, and rabbit manure are considered cold and can be added to worm bidding without composting. They don't produce a lot heat.

The manure will be used to provide food and bedding for the worms. To get the best results, you should mix 50% of manure with half of paper. You can use fresh manure to generate heat, which is okay during the colder months. Aged manure should be sat for six month to prevent heat accumulation. You should not use hot manure in your bins. It's too hot, and can raise the temperature too much. Hot composting is not great for gardens.

Wood Chips

It is vital to limit the use of wood chips in bedding. Hardwood can be used as bulk, and also to increase the airflow through the bin. While wood chips can take years for to completely decompose but will still provide food and shelter for any fungal growth, it is possible to make wood chips.

It was an accident that I learned that worms like fresh rabbit manure. Rabbit manure can be used fresh or preserved for use in worm bins and gardens. There is an increase in the number and quality of worms everywhere we keep our rabbits.

Make sure your bedding is damp when adding it to your worm boxes. I don't mean damp, but damp like a sponge that was wrung out. Fluff up the bedding once in awhile to prevent it becoming too compacted and smelly. As I mentioned, you should also remember to feed the Worms in various places.

Specifications of suitable bedding material

You must ensure you use the right types of bedding materials when you first set up your worm farms. Here are some things you should include in your worms bedding

It should be of neutral PH.

The worm's skin is sensitive and should not be exposed to any abrasive substances or sharp objects.

It must retain moisture.

It should allow airflow.

It should be odorless.

It is vital that your bedding is free from chemicals.

It blocks out the light.

Your worms will eat bedding. Bedding could account for up to half of a worm's daily diet. It is thought that brown cardboard can be used as bedding or food for worms. This is especially true if they are red. Because it allows air to flow better, it is easier for your larvae to consume.

It doesn't matter what type of bedding you choose, moisten it with 60% to 60% moisture. It is much simpler to add water than to get rid of it.

Bedding Ratio

A 50-50 mixture of cococoir and shredded papers is a great base for a new compost bin. A second option is to include 25% shredded leaves or 25 percent manure or mature soil, 25% shredded paper and 25% coco-coir. Also, add a light dusting with grit.

Remember that it can take several days for the microbial communities to reach the size they need to start to digest food waste. For the best environment, ensure you prepare bedding materials several weeks in advance so that they are ready when the worms arrive.

You can increase microbial activity in bedding materials by adding 1 cup of wheat flour, cornmeal, or oat. Inoculate your bins to encourage rapid decomposition. Blend a few tablespoons of the mixture with the bedding material to add a little more to the bottom.

Beware of Toxins

It is vital to keep toxins from reaching your worm bins. It's not hard to believe that we are constantly being exposed to unknown toxicities. It is important to keep toxic materials from your worms' worm bin. Do not give your worms these substances!

Let's discuss toxins

According to the Oxford dictionary, toxins can be defined as "antigenic poisons/venoms of plant and animal origin, particularly those produced or obtained from microorganisms" that cause disease when present in low concentrations.

Garden Toxins

There are many things that can be poisonous in the outdoors. However, ALL plants contain toxic substances. These toxins don't always pose a threat to humans or animals. However, they can protect the plants against herbivores who are looking for food. You might not know it, but the toxin may remain in the dead matter even after the plants have died.

An earthworm's gut may contain certain molecules that neutralize the effects of polyphenols. These compounds give plants color, act as antioxidants and deter herbivores. There are some natural plant compounds that can be used to keep worms from getting away. These are the toxins that worm farmers must avoid!

Take into consideration the common members the nightshade group, which include tomatoes, eggplants, tomatoes and peppers. This family includes plants that produce saponins as well solanine and atropine-like chemicals. This is why we try not to feed our worms certain foods. The toxin in the Chrysanthemum flower petals is called pyrethrin. This kills all insects on contact and also makes it difficult for worms to survive.

The following garden plants can also be dangerous for worms: the leaves and stems of the neep trees, onions, hot peppers as well citrus peels and eucalyptus. The best way to control these plants is to grind them and then add water. If you decide to make this a reality, ensure that you wash everything you are going to be giving the worms.

Man-Made Toxins To Avoid

The environment can be affected by pesticides, fertilizers for plants, and mosquito fog. Caffeine and theobromine can make natural cocoa bean mulch dangerous to worms. They're safer than chemical insecticides because they're natural, and therefore they're less toxic to worms than chemical pesticides.

We regularly consume and throw out toxic products. This can disrupt the natural balance of nature. These harmful compounds can be transported miles away by rain and wind, which introduces them into the ecosystem.

Toxic Kitchen Scraps

I know that I already listed foods that should be avoided when feeding your worms. But here's a deeper look. Our leftovers are delicious but not all of them can be used for vermicomposting. Salt, acid and fermenting foods can cause harm and death to your worm's environment. They may move to another place in a natural setting but they won't be allowed to do so if they're trapped in a container.

Salt can damage the delicate skin of your worms and cause them to die. Although it takes high salt concentrations to have such a powerful effect, you cannot be certain how much salt is added to your food. Avoid putting salty food in the bin. It will only take a few months before a wormbin becomes too salty. This can occur with natural salts found in food as well as any salts added to food as flavor enhancers and preservatives.

The skin of your worms can be also burned by acidic food, which is why they are not allowed to go into the compost. People with food allergies can sometimes eat a bit of something they shouldn't. We don't always have the ability to judge what was eaten. It's not about what they'll eat, but how it might affect their health.

Overfeeding worms causes stench and insects as well as problems. The food becomes fermented and toxic gasses build up when your worms don't want to eat it all. It's like living with carbon monoxide gas leaks in your house. It's not something you can see or smell, but it's there.

Tips and Tricks for WormHealth

All of this information may seem scary, and perhaps even impossible to avoid, but it is not so bad. Your worms will be happy and healthy if you are careful.

The conditions in your worm bin will greatly depend upon the food source and water that you use, as well how often you collect castings. To keep everything in good shape, I have some reliable safeguards.

Remove Offenders

If something smells bad it's probably because of the vermicomposter. Take a look at the vermicomposter, and get rid of any that is causing the stench. If it is present, you should move on to the periodic pourover.

Rinse & Repeat

Even though wet food may seem gross, this is not the case for worms. You should rinse your worms out of any leftover food to get rid any potential harmful ingredients. Plain is the best choice.

Periodic Pour Over

The food should be placed in the bin. However, you can give the bedding and the food a nice wash. It is important to rinse out any salts or sludge that has accumulated. It is vital that the worm containers have adequate drainage in order to allow this process to work. Without proper drainage, your worms are doomed to die.

Low & Slow wins The Race

To ensure that you get the best results, always give small quantities of any new food to the worms. Start by placing a little in a corner. Then, observe how the worms react. You can add more once you are certain that there are no adverse reactions.

Use a dual-probe meter

A dual probe meter measures pH levels to help you maintain controlled acidity. The moisture meter can be used to tell you if you should add more water. It can be difficult looking in a bin to determine what is going on.

Always do your research! You should always do your research before you put any food items in your bin. Although this list is not complete, there are so much more foods than I can remember.

Chapter 20: What to do and what not to do when you're involved in Worm Farming

There is so much information out there about worm farms, it can be overwhelming. But don't panic. Although learning anything new can be overwhelming, once you begin to put what you have learned into practice, your mind will become more organized.

Placement and location of Worm Bins

A plastic tub with worms left outside in the cold will result in their death. Although they are capable of digging deeper and digging for warmth, worms in a bin can only go so far. Do not leave your bins out on the ground in winter. This will cause them to feel the cold radiating from the ground.

Worms are most vulnerable in winter and the summer. Your worms could become dehydrated if they are exposed to too much heat. If it is too cold, the worms will freeze. It's important to monitor outdoor worm bins when it rains because excess water can cause worms to drown.

Your worm farm should be indoors so you don't have to worry. That depends on where your worms live. If they're in an unheated area such as a garage or basement, they might require some heat source or insulation. Your worms should be indoors during summer so they don't get too hot or in direct sunlight. Let's talk about the seasons.

Spring

The rain comes with spring weather. Remember the cooler nighttime temperatures that occur in early spring. Make sure your worm farms are raised above the ground for proper drainage during rainy season. Because the bins absorb the water in the puddles, it is especially important that they are not on the ground when it rains or gets really wet.

As it's not too warm outside, your worms can be at their best in a sunny spot. There is no need to worry about rain getting into the worm bins' tops. They have a tight seal that should be free from cracks and holes. You can place them under an umbrella or cover them with a tarp if it is raining heavily.

You can bring them in if temperatures fall below the safe zones for worms or you can make sure that they have some insulation. Even though it's not absolutely necessary, it may be beneficial for your health and the comfort of the worms.

Summer

Too much sunlight and too warm temperatures are bad for worms. I am sure you have seen dried-up insects on the pavement or in a parking area at some time in your entire life. Remember that worms can only survive if there is enough moisture.

Your worm farm should be located outside during the hotter months of summer. You should never leave your bins out in direct sunlight. Be aware that vermicomposting generates heat. If you add the heat from the summer sun, it will be dangerous.

During the summer, you should check the worms' bedding at least twice per day. When it's hot, you'll need more water. But don't forget that too much water is also bad. To test if the bedding is damp, press down on it. In the heat of summer, cool and shaded spaces are important.

Autumn

After the spring rainy period, Autumn brings us the rainy season! The spring rainy season is over, and the autumn will see the rainy season. Keep the bins warm enough during colder nights and cool enough on warmer days. You can always place a thermometer into the bins so you can see how hot it is. You cannot rely on the temperature outside because it will be different inside.

Winter

If you live in an area with low temperatures, it will be important to insulate your outdoor wormbins. If you live in North Dakota's winter, you will want to bring your worm bins in. Remember that worms only can go so far in a bin if it's cold. If you're cold your worms will be too.

Common Mistakes When Vermicomposting

Vermicomposting is made easier by using worms. It is normal for new farmers to make mistakes. To help you be aware of common mistakes in worm farming, I'll go through them. It would be a shame not to invest your time, money, and effort only to have it all go to waste. Okay, maybe mistakes aren't always so terrible, but there is still chaos.

First Mistake Overfeeding Your Worms

I really don't mean that I sound like a broken record. But I want you all to be successful so I'm repeating the words. Overfeeding worms can be a problem in many ways. The theory is that worms can consume all of their food scraps each day. But there is always a chance that it might be less. The amount of worms consumed varies based on temperature and other factors.

A fool-proof strategy is to feed your pets every 2 or 3 days. You should not add too much food at once. Through time and experience you'll be able to get an idea of how much yourworms will eat. Before adding any more, ensure they have eaten what you've already given them. In a matter of weeks, an entire feeding should be finished. It should be gone within a week or two.

2. Mistake #2: Feeding Your Children the Wrong Food

Worms require a healthy and balanced diet. They need small, regular meals. The longer it takes to decompose food scraps, the greater their size. You can risk the entire bin being spoiled by meat scraps and processed food. You cannot eat non-food products.

You should only feed your worms non-acidic vegetable scraps and fruit scraps. These include bread, coffee grounds as well as grains, tea bags, biodegradable teabags without staple and string, and pasta. Crushed eggshells are a great way to provide calcium and grit. If you have old grass clippings and/or animal manure, don't forget to add them.

3. Mistake #3: Too Dry Or Too Wet

Too much water in a worm box can cause them to drown. Too dry bedding can cause the worms to lose their ability to breathe and will not be able tunnel well enough. As I mentioned, take a small amount of the bedding and squeeze. If water drips from the bedding, it's probably too wet. The bedding for worms should feel like a damp sponge that's been wrung.

No Harvesting Worm Castings - Mistake #4

If you are one of those people who practice vermicomposting in search of the "black diamond", it may be easy to forget to harvest your castings. We are excitedly awaiting the harvesting days. This is a crucial aspect of worm farming, but it's easy to forget. You can't leave a cat without using their litter box. For worms, the litter box is a trap they can't escape from, so it will be crucial to remove the castings as often as possible. You don't have to worry about harvesting, as we will discuss it in detail later.

Mistake #5 Temperatures too hot or too cold

If the temperature drops below 54° Fahrenheit, your larvae will be less productive. It is normal.

Your worms are still safe. If the temperature drops below freezing, your little wigglers will die. Your worms can be slowly cooked if the temperature is higher than 84 degrees Fahrenheit.

You will make mistakes. That's okay. Keep in mind that these worms live creatures. Be aware of potential dangers and take steps to avoid them.

How to Keep Worm Bins Clean

There are some things that you should look out for after your wormbed or bin has been installed and established.

The worm bin beds must be constantly mixing and settling to provide a healthy environment where they can thrive.

The worms should be visiting the food pockets. Just dig gently around the area you've placed the food scraps. There should be worms in that area.

If you find other insects in your bin, don't panic. A wormbin will contain a complex food web, which creates what should be a well-balanced environment. The more productive their system is, the healthier their bin environment. It's not necessary to worry because all of those insects love the inside of the bin. If your worms are tempted to wander off after arriving, you can use an overhead lamp. They won't go near the light, so make sure to position the light in a way that they don't mind.

The bedding should absorb some moisture from the food scraps. You should not add water as soon as possible. It is better to put the veggie scraps through a processor or blender before you add them to the container. This will ensure that the worms have enough moisture to be able to eat the vegetables and digest them.

The number of organisms and diversity in the wormbin will increase over time. There is no reason to panic; the microorganisms within the bin won't attempt escape. They are provided with all the necessary nutrients to survive if they are taken care of properly. They can live happily in their own home if they have a dry environment and are fed regularly.

Signs of a Healthy Bin Ecosystem

Many organisms are busy in the worm box. They are vital for the ecosystem of composting. Your worms depend on this food web for breaking down large food particles and organic matter into smaller bits that are suitable for them. All kinds of bacteria, fungi, protozoa, and roly poly pill bugs will be found. Do not be surprised to see a few spiders.

Organisms

These organisms will multiply, which is a good sign for a healthy bin and a productive decomposition process.

Springtails, Collembola family: Springtails may be white, grey, or brown in color. They have six legs, three parts of the body, and two antennae. These organisms are named Springtail because they have a spring organ to propel them forward.

Pill Bugs (Isopoda), also known as Sow or Pill Bugs, are often called "roly-poly" due to the way they roll up when they feel threatened. These little guys will eat any material that is high in lignins or cellulose. People remember their armored shell that is segmented and has seven pairs, as well as two antennae.

Enchytraeidae potworms: These threadlike worms are frequently mistaken for redworms. They are also called white worms by some because they are white. This is the perfect place for these little creatures to live. The more you have white worms, the better. If you see a lot of white worms, it could be a sign your environment is too dry. To absorb excess moisture, you can add more bedding and shredded paper. To create a balance, you can fluff up the bedding.

Acarina Mites: These creatures are usually the most common and easily visible. They feed on organic matter and fungi as well as other organisms. You will find them under the worm bedding.

Millipedes (Diplopoda: These long, slow-moving, insect-like creatures can be found in small numbers all over a bin. They are long and segmented. There are two sets of legs for every segment and two antennas.

Centipedes - Chilopoda (Chilopoda). They look very similar to millipedes. But they have only one leg on the majority of their bodies. Large pincers. They are usually reddish in color. They are quick and will often make it into the bin. If you spot one of these creatures, take them out immediately. You can pinch them!

Diptera fruit flies: These little annoying insects are an important part of the worm-bin community. They have larvae which are voraciously decomposers. Fruit flies attract to the acids in the vegetative matter that is decaying.

Bacteria

Bacteria are the largest number of organisms found in a bin with worms. They, and molds as well as fungi, are the most abundant organisms in a worm bin. Each of them eats organic matter that is decaying and makes enzymes to help break down and simplify it.

It is safe to say that your worms' bins should function as an ecosystem or productive environment. You can take good care and consider other living organisms in order to improve the environment for your worm.

Chapter 21: Worm bin care tips

Worm bins are often difficult to maintain. However, these tips can help to maintain a healthy and productive environment for composting.

Regularly Monitor the Bin's Water Content

Some manufacturers of worm boxes will advise that you add water in your worm containers to increase the production. However, adding water to your bin should only be done if it's large enough to feel dry to the touch. Leachate is the liquid that drains from the wormbin. If it's not diluted, it can be very concentrated and cause severe damage to plants.

If you see a lot of leachate running out, it means that the environment has become too wet. To increase oxygen flow, add dry bedding. Every few weeks, you should raise the dry bedding off the top layer of your bin and gently dig down into the tray. You want to disrupt as few worms as possible.

You should inspect the tray and determine what the material is like at the bottom. If the tray looks wet or is very smelly, that means the bins require more airflow. To allow any liquid to drain out, make any necessary repairs to the drainage system.

When you feed your worms or check on them, be sure to take a good look inside. It could be that you've just recently added some green food scraps or the bin has become anaerobic. To dry it, you'll need the bedding lifted up and the container let air flow through.

Temperature & Weather Regulation

Worm farming does not mean that you can simply "set it up and forget about it". It is important to keep your worms healthy and well-managed. Your worms will thrive in an environment that maintains a constant temperature. The temperature should be between 0 and 41 degrees Fahrenheit. Direct sunlight is dangerous and it's a bad idea to place in direct sunlight.

Predators

They don't simply say "the earliest bird catches the fish" without good reason. Like moles and birds, worms are delicious for birds. If you have worm bins outside, keep an eye on them for signs they might be hunting. Mice will make nests by chewing holes in worm bins if they have access to them. Make sure to check the bin every time you feed the worms, or make sure the worms are not trying to get into it.

Amendments or "Buffers".

To improve the environment, some farmers add extra ingredients. Understanding how the ingredients work together in an environment is essential when using amendments. A system that has too much of an element can lead to imbalance. These additives include oystershells, lime and eggshells as well as rock powder.

The technical term pH/acidity/buffer gives rise to "buffer". It is sometimes also called an "acidity regulator". To keep the worm farm's pH from rising, certain amendments can be made. These ingredients not only provide a buffering effect, but also provide a rich source of grit as well as calcium and minerals for the larvae.

Acidic conditions can cause worm farms to become acidic. Ammonia will form and the gas will remain in the bin. This can cause protein poisoning. If the environment is acidic, your worms may become sick, unwell, or even die.

We have already talked about how worms like chickens need the grit in order to grind their food. Buffers provide vital minerals for their health. However, they could be lacking if they are locked in a box. Calcium is critical for the worm's cocoons, as well their calciferous gills.

Worm bin buffers:

Garden Lime

Zeolite

Dolomite

Azomite

Oyster Shell flour

Finely Ground Eggshells

To add buffers to your bin, sprinkle some only once it has been started. You can then sprinkle the buffers over the food every time the worms are fed. You may find the right product at a hardware or garden supply store.

I've also heard that wood ash can be used as a buffer in a "worm bin". Ashes should be avoided. You will increase the acidity or pH of the bins, in addition to any chemical compounds. High pH levels can prove to be as harmful as high acidic levels.

Zeolite & Azomite

The volcanic rocks azomite, and zeolite are both volcanic rocks. They were formed millions years ago. The mines from which they were extracted will determine the mineral makeup of each one.

Zeolite is used to balance pH and to provide grit. Azomite is a great mineral supplement for the worms. Zeolite makes good grit when they are crushed small enough.

Use Hydrated or Slaked Lemon

Your worms can be killed with slaked lime or hydrated lime. You must use garden lime, or dolomite. Garden lime is powdered chalkstone or limestone. Dolomite can be described as garden lime with more magnesium.

Hydrated lime/slakedlime is made from limestone heated in a furnace until it forms a powder. Heat is used to transform calcium carbonate from calcium hydroxide. It's not something our worms require.

Chapter 22: Troubleshooting

You've seen many common problems associated with worm farming. Let's now talk about how we can troubleshoot these problems and solve them.

Insects & Fruit Flies

While fruit flies can sometimes be found in your worm boxes, they can cause problems if they grow in numbers and start escaping. Centipedes and other insects can eat worms. You can safely remove any red mites that you find and place them in a different area. Red mites can be a nuisance, but increasing numbers increases the risk. It is clear that the main problem is an excessive number of unwelcome guests in your home. These pests can easily be controlled using the techniques below.

You should avoid putting rotten or rotten food in your worm boxes. Decomposing foods are more likely than other food to attract fruit flies.

You can cut food into smaller pieces or throw them in a food processor because different insects eat food at different stages. For worms, smaller bits of food mean that they can eat quicker and are less likely to be invasive.

Don't feed too many worms. A warm worm bin with plenty of food will attract more insects, and perhaps even smaller animals.

To minimize the pests' smell and exposure, place the worms' food in a container.

Keep the bin free of food rotting. It is important to keep an eye on your worms' population and numbers. If you can spot any potential overpopulation or crowding, it will be easier for you and your bin to stay in better shape.

Take a look at the food and remove any rotten foods. To ensure that the babies hatch, fruit flies will lay their eggs on decaying food. This is usually a preventative measure that also serves as an initial step towards tackling the problem.

Flypaper strips can also be taped or stapled on the lid. To prevent new insects entering, you can hold a flypaper piece up to the bin. If insects fly or crawl across the cover, the strips on its inside will catch them. Flypaper and flytape can be purchased at low prices in most hardware and grocery stores.

A small bowl containing apple cider vinegar and a bit of dish liquid can be used to create a fly trap. You can place the bowl next to your worm boxes and it will attract and kill the insects. Keep the fly trap liquid fresh by changing it often to attract unwanted insects.

You can place a complete sheet of newspaper on top, almost like a blanket. Fruit slides and gnats will congregate on the newspaper for some reason so it is important to change the newspaper often.

Many insects, including fruit flies, are attracted by an acidic environment. This is a sign that the environment is acidic. Acidic spaces and Worms don't get along. You can neutralize the acidic conditions by adding a little lime to your bin.

If you find fruit flies, gnats or other pests in your bin, it is best to take it outside. You can let the bin air-out for a few hours but make sure it's not exposed to direct sunlight.

If the problem persists you may need a new bin made from scratch. A soil test kit can provide insight into certain aspects of the environment and help you to further troubleshoot.

You can safely take any harmful insects out of the bin and dispose them outside. An excess of any kind of insect means that there is something wrong with the environment. Find the excess of critters, identify what attracts them, and then limit or eliminate the problem.

Odor Problems

Our nose will tell us when something is not right in our worm box. A healthy worm box should smell earthy, just like fresh dirt and the forest after a good rainfall. If you start to notice foul or unpleasant odors inside your container, this is a sign of something that is not right.

Hot compost can often smell unpleasant. Vermicomposting has the ability to reduce unpleasant odors, which is why both farmers and gardeners love it. A worm farm ensures that food is processed much faster than it would if it weren't for the worms. The bin should not smell if it has a lot of worms. If your bin smells musty or funky, then you need to identify the source.

Is this the reason for the smelly food? Look around in your feed pockets for the cause. Take a good look. You can remove large quantities of undigested organic matter. Overfeeding the worms leads to too many worms eating food that should not be there. There is also the possibility of some 'wrong' foods being thrown in. Rotting milk or meat can lead to some very strange results.

Is the worm-bin too wet Give your bedding some pressure and take a look at the situation. If the bedding is more wet than a sponge, you need to lower its moisture. To make sure that your drainage holes don't get clogged, check them. To help improve the airflow, fluff your bedding. Next, add a layer with fresh, dry bedding to absorb moisture. If the bin is not dry enough, you may need to build a new container.

Is your bin drying out too much? Keep in mind that worms don't need oxygen to breathe because they don't possess lungs. You can inspect your bins to see if they are dry. If they are, you could rinse the food scraps off or use a food processing machine to make the food more liquid. This adds moisture without having to water the bins. You can sprinkle dechlorinated salt on top if it's too dry. You can lightly water the top, and then wait for it to dry for five minutes. Add a little more water and continue fluffing the bedding to create a moist environment.

Is the bin pH balanced? Healthy living conditions for worms requires a pH 7 or higher. Test your bins periodically with a pH probe, or soil testing kits. Alkaline and acidic foods can alter the pH levels and may require some adjustments.

Worm Death

It is vital to keep an eye on the worm population at your farm. You might see a rise in numbers, which is great. But if you notice a decrease in numbers, there is something wrong. Worms do not disappear. You may notice that a few worms are missing. Imagine your living worms crawling across the bin. You should assess the worm bin immediately if you see them trying to escape. If you are able to identify the root cause of the problem sooner, it can be fixed.

The bin has become too saturated and the worms have begun to drown.

Your bin is too dry. Worms are drying out or dehydrating.

Airflow is not sufficient to prevent toxins from pooling and puddling up.

Without enough moisture, the worms will suffocate.

Worms have a hard time finding enough food. This can lead to them eating their own castings. Castings that they consume are toxic to them.

Extreme temperatures may be exposed.

Dead worms quickly decay. The whole farm could end up dying if you aren't paying enough attention.

Too Many Suppers Left on the Plate

This is a simple solution. If you see a lot left in your bin, this is an indication that you're feeding them too much food. It can be difficult for the worms and other insects to find the food by leaving it directly on the surfaces. It is better to hide the food a few inches beneath the bedding. This makes it easier to find. Allow them to eat small portions every few weeks. They should not be given food leftover from a week or so ago.

Critters

Indoor worm bins won't be an issue. But that doesn't mean they aren't possible. Rodents, raccoons or birds might steal your lunch box

..... I am referring to the worm box, which can be used for tasty snacks. They like leftover food just as much the worms. Some of them even enjoy nibbling on worms!

These creatures are attracted to strong odors like oils, fish or meat. This is one of many reasons these food products shouldn't be thrown away. They won't just eat them, though.

For food odor control, ensure you are burying any leftovers.

You should ensure that your bins have tight fitting lids. You can secure your lids with bungee ropes, bricks, and heavy rocks.

To ensure safety, repair any damage caused by the critters and consider moving them elsewhere.

Different animals may leave clues which will help you identify the person who has gotten into the honey pot. Mice will chew through the container to search for food, and then burrow into a hollow. Raccoons could knock the whole bin over; it all depends what the critter is trying to do.

Escaping Worms

When your worms live in a happy environment, are healthy, and feel comfortable, they will not go anywhere except to the food pockets. These are the things to look out for if your worms seem to be trying to climb out from the bins.

Something is off balance in the bin. You will need to identify the problem and solve it.

You might find your worms trying to escape if you recently moved them, especially in a car. It could cause an uncomfortable vibration under theworms that can make them want to escape potential danger.

Adult worms can sometimes escape after the baby worms hatch. You can ask them to stay in the bin so that the population balances itself. This cue could also be used to signal that it is time to create a second worm container to expand your farm.

Temperatures: Too cold or too hot

The ideal operating temperatures for worm bins are between 57 degrees Fahrenheit, and up to the 1980s/'90s. The bin temperature should never drop below 40°F. Worms will freeze and eventually die if this happens. Temperatures in excess of the mid-'90s could cause your worms and other animals to become sick.

Where are The Babies?

You can have happy and healthy, reproducing worms if they're well taken care of. If they are happy and healthy, they will lay egg-filled eggs. If the temperatures are perfect, the eggs should hatch in one month. If you haven't seen any babies in the container, it is worth giving it a while. Baby worms will emerge once everything is in order. It's six weeks later and there are still no babyworms. It is time to assess the environment and make necessary changes.

As you get more experience, troubleshooting issues and correcting them will become second nature. You'll gain more knowledge and skills as you work on your worm farm. You don't have to be an expert farmer or know enough about worm care.

www.ingramcontent.com/pod-product-compliance
Lightning Source LLC
Chambersburg PA
CBHW050411120526
44590CB00015B/1928